AF616702

THE EFFECTS OF AGING AND ENVIRONMENT ON VISION

THE EFFECTS OF AGING AND ENVIRONMENT ON VISION

Edited by

Donald Armstrong

State University of New York
Buffalo, New York

Michael F. Marmor

Stanford University Medical Center
Stanford, California

and

J. Mark Ordy

Fisons Pharmaceuticals
Rochester, New York

PLENUM PRESS • NEW YORK AND LONDON

Library of Congress Cataloging-in-Publication Data

American Aging Association Symposium on Environmental Design for Optimum Vision in the Elderly (1985 : San Francisco, Calif.)
The effects of aging and environment on vision / edited by Donald Armstrong, Michael F. Marmor, and J. Mark Ordy.
p. cm.
"Proceedings of an American Aging Association Symposium on Environmental Design for Optimum Vision in the Elderly, held October 11-12, 1985, in San Francisco, California"--T.p. verso.
Core content comprised of conference papers, many updated; includes some new papers.
Includes bibliographical references and index.
ISBN 0-306-43920-4
1. Vision disorders in old age--Congresses. 2. Vision disorders in old age--Environmental aspects--Congresses. 3. Eye--Aging--Congresses. I. Armstrong, D. II. Marmor, Michael F., 1941- III. Ordy, J. Mark. IV. Title.
[DNLM: 1. Aging--congresses. 2. Environmental Exposure--congresses. 3. Vision--in old age--congresses. 4. Vision Disorders--etiology--congresses. 5. Vision Disorders--in old age--congresses. WW 140 A512c 1985]
RE48.2.A5A44 1985
618.97'77--dc20
DNLM/DLC
for Library of Congress 91-24104
CIP

Proceedings of an American Aging Association Symposium on Environmental Design for Optimum Vision in the Elderly, held October 11-12, 1985, in San Francisco, California

ISBN 0-306-43920-4

A Division of Plenum Publishing Corporation
233 Spring Street, New York, N.Y. 10013

Printed in the United States of America

PREFACE

This book derives from a symposium conducted in San Francisco CA, entitled "Environmental Design for Optimum Vision in the Elderly" that was sponsored by the American Aging Association, October 11-12, 1985. Presentations from this symposium comprise the core content of this volume. However, manuscripts have also been accepted from additional authors whose topics and research findings increase the scope and goals of this volume. Previously unpublished data is found in several of the chapters. In addition new data and references have been incorporated at the end of chapters in order to provide a current update of the subject.

The broad aims of the papers in this volume were to examine the effects of various environmental factors, long-term occupational hazards, and toxins on basic visual functions in relation to physiological, biochemical, morphological, and pathological alterations in the eye and visual pathways, and centers of the brain. As part of the more specific aims of this volume, the editors have provided the following framework for the specific topics included in this volume: I) Epidemiology, Clinical and Psychophysical Research, II) Ophthalmological, Biochemical, Physiological and Anatomical Studies, and III) Environmental Hazards.

Although the visual system is one of the most extensively studied and better understood components of the brain in neuroscience, studies of vision, aging, and environmental interactions are only in an early stage of development. The number and proportion of elderly in the total population are increasing dramatically, yet the loss of visual functions and increases in ocular pathology that often accompany aging are not fully understood. Because a multidisciplinary neuroscience approach to vision and aging includes molecular, physiological, biochemical, behavioral, ophthalmological, and epidemiological studies, the topics covered in this volume should enhance interdisciplinary communications among scientists, clinicians, and individuals engaged in the delivery of health care services. It is anticipated that future research on the topics covered in this volume may result in a better understanding of vision and aging and also stimulate progress towards development of improved visual environments for the elderly.

The editors gratefully acknowledge the editorial assistance provided by Mrs. L. Kaite and Mrs. P. Pagano.

D. Armstrong
M.F. Marmor
J.M. Ordy

CONTENTS

III

ENVIRONMENTAL HAZARDS

VISUAL ACUITY, AGING, AND ENVIRONMENTAL INTERACTIONS:

A NEUROSCIENCE PERSPECTIVE

J.M. Ordy[1], T.M. Wengenack[2], and W.P. Dunlap[3]

[1]Fisons Pharmaceuticals, Rochester, NY,
[2]Univ. of Rochester, Rochester, NY, and
[3]Tulane Univ., New Orleans, LA

INDIVIDUAL, SOCIAL, AND SCIENTIFIC PERSPECTIVES ON VISION AND AGING

Individual experiences, social, and scientific observations in many disciplines have noted that: 1) vision is of great importance for acquiring detailed information about physical and social environments, and 2) that impairments of visual functions and the incidence of ocular pathologies appear closely associated with advancing age (Birren and Williams, 1982). Although cross-sectional studies may overestimate age declines in various visual functions (Storandt, 1982), they have suggested that impairments in such functions as acuity, contrast sensitivity, adaptation, stereopsis (Pitts, 1982), color vision, glare sensitivity, field of vision, accommodation (Carter, 1982), and ocular motility (Leigh, 1982) may occur with increasing senile ocular changes (Weale, 1982) as part of normal aging. In contrast to the explosive research on visual development, research on vision and aging is still fragmentary and in an exploratory stage (Ordy and Brizzee, 1979; Birren and Williams, 1982). Major unresolved problems and challenges in epidemiological, clinical, and basic research on vision and aging include: 1) distinctions between effects of normal aging and pathology, 2) validity and reliability of conclusions about the adverse effects of age on vision, based on methodological differences between "age-changes" reported in longitudinal, and "age-differences", reported in cross-sectional studies, and 3) reality and magnitude of mechanisms of repair, recovery, redundancy, or so-called "neural plasticity" in the visual system from maturity to senescence. Research on all of these aspects of vision and aging would provide critically needed information on compensatory mechanisms, response strategies, or novel environmental designs, for optimizing vision in the elderly.

NEUROSCIENCE PERSPECTIVES ON VISION AND AGING

Earlier reviews of vision and aging in man and nonhuman primates have examined age differences, or changes in such visual functions as acuity, contrast sensitivity, brightness sensitivity, color discrimination, and stereopsis in relation to some physiological, biochemical, and morphological alterations in the retina, and in the visual pathways and centers of the brain (Ordy and Brizzee, 1979; Ordy, et al., 1982). Because of the enormous research progress in vision, particularly during development, a contemporary neuroscience perspective on vision and aging has also expanded to include

The Effects of Aging and Environment on Vision
Edited by D.A. Armstrong *et al.*, Plenum Press, New York, 1991

multi-disciplinary approaches extending from: 1) molecular events in conversion of light energy into neural signals, or photoreceptor transduction of photons by rod/cone pigments and G-proteins (guanine-nucleotide-binding-proteins) into graded potentials, from depolarized dark currents (Hubbel and Bownds, 1979; Nathans, 1987), 2) transformation of an optical image on the retina into the "language" of neurotransmitters of neurons, as pattern of "on-off" electrical activity (Kaneko, 1979; Daw, et al., 1989), 3) retinotopic and geniculostriate organization for parallel visual information processing (Rodieck, 1979), 4) functional mapping of ocular dominance and orientation selectivity in visual areas of the cerebral cortex (Van Essen, 1979) behavioral discrimination or visual psychophysics (Westheimer, 1984), 5) age-related macular degeneration (AMD)(Lovie-Kitchin and Bowman, 1985), and 6) to inherited or acquired visual impairments and ocular anomalies (Foster, 1990; Kulikowski, et al., 1990).

It is widely recognized that the mosaic, or geometry of the cone and rod photoreceptors, particularly in the fovea of the retina, provides the spatial information available for the retinotopic, and geniculostriate stages of visual spatial information processing. It has become evident that photoreceptor inner segment diameter, spacing, and the discrete geometric sampling "array" may comprise the basic anatomical bases for acuity, spatial discrimination, and pattern recognition (Hirsch and Curcio, 1989). Recent quantitative studies have begun to examine foveal cone and rod densities in the human (Curcio, et al., 1990), and in the rhesus monkey retina (Schein, 1988). Visual acuity and foveal cone/rod density have been examined in the retina of the aged rhesus monkey (Ordy, et al., 1980). Correlations have been reported among loss of photoreceptors, lipofuscin accumulation in retinal pigment epithelium (RPE), and macular degeneration with age in the human retina (Dorey, et al., 1989). Several studies have reported preliminary quantitative data on possible age differences in foveal cone/rod densities in the retina of young, middle aged, and old human subjects (Youdelis and Hendrickson, 1986; Gao, et al., 1990). It seems likely that psychophysical and morphometric studies with humans and nonhuman primates will begin to provide clarification of such fundamental problems in vision and aging as age differences , or changes in visual acuity, foveal cone/rod density, lipofuscin accumulation in RPE, and cell loss in the geniculostriate system in human subjects, and in nonhuman primates (Ordy, et al., 1991).

VISION, DEVELOPMENT, AGING, ENVIRONMENTAL INTERACTIONS

It is generally not recognized that in single-, or multi-disciplinary studies of the effects of age on vision, it is very important, but difficult to distinguish the events that are genetically programmed, or intrinsic to aging, from effects that are attributable to various environmental factors, such as exposure to environmental light hazards, toxins, nutrition, health status, and life history (Storandt, 1982). In the context of vision and environmental interactions, it is now generally recognized that the environment may play a critical role in the development of the mammalian visual system. A large number of studies during the past decades on "neural plasticity" in the visual system during development have demonstrated that such alterations in rearing condition as changes in activity levels, exposure to changes in light-dark cycles, and "sculptured visual environments" can lead to modifications of neural connections in the visual areas of the mammalian brain (Hirsch and Hylton, 1984; Teller and Movshon, 1986). The developmental programs of the visual system of mammals appear to depend in part on interactions with the visual environment for their completion (Wiesel, 1982). To what extent, if any, a "neural plasticity" exists, in the mammalian visual system from development, to maturity, and to senescence, has become of considerable interest (Spear, 1985; Lund, et al.,

1988). In the context of aging and recovery of function after injury in the mammalian central nervous system (Scheff, 1984), studies have shown changes in occipital cortex neuron and glia numbers, as well as in dendritic branching and density, in response to differential visual environments and levels of activity in the rat during aging (Diamond and Conner, 1984). It is generally recognized that visual impairments following visual system damage in infants are less severe than the impairments following similar damage in adults (Spear, 1985). It has been proposed, however, that loss of neurons during aging may result in dendritic proliferation in "surviving neighbor neurons" as a compensatory mechanism of repair or "neural plasticity" (Coleman and Buell, 1985). As yet, it remains to be determined if the phenomena of morphological "neural plasticity", result in recovery of specific visual functions in the mammalian nervous system from maturity to senescence (Spear, 1985). As a broad generalization, it has been proposed that age-related declines in "neural plasticity" may some how be associated with declines in visual and cognitive capacities, including learning and memory, in normal aging (Petit and Ivy, 1988), with diminished "neural plasticity" occurring in such age-related neurodegenerative disorders as Alzheimer's disease (Flood and Coleman, 1990).

PSYCHOPHYSICAL, NEURAL, AND EPIDEMIOLOGICAL DISTINCTIONS IN VISION AND AGING BETWEEN NORMAL AGING AND PATHOLOGY

Several major and related challenges in psychophysical, clinical, and epidemiological research on vision and aging include: 1) defining, or differentiating normal aging from age-related diseases, 2) measuring the onset and magnitude of the effects of normal aging on visual functions in longitudinal and cross-sectional designs, and 3) estimating the incidence and prevalence of visual pathologies with increasing age in epidemiological studies. Both the number and proportion of the elderly in the USA and other industrialized societies are increasing dramatically (Siegel, 1980). Although many elderly do not experience severe visual impairments, declines in visual functions do occur in individuals who live long enough (Carter, 1982; Pitts, 1982) and the incidence of ocular pathology increases markedly with advanced age (Greenberg and Branch, 1982). Because visual impairments not only increase as a function of aging, but may also coexist with other sensory and physical impairments (Kirschner and Peterson, 1979), it has often been assumed that visual impairments in aging may be part of a disease process (Birren and Williams, 1982; Storandt, 1982).

Comparing the onset and magnitude of age differences, or changes in such basic visual functions as acuity, color vision, and stereopsis during normal aging has also remained a methodological challenge because the estimates depend upon the extensive use of cross-sectional, rather than longitudinal designs (Storandt, 1982). In cross-sectional designs, subjects

Table 1

Cross-Sectional Design		Longitudinal Design*	
Year of Birth**	Age	Time of Measurement	Age
1910	80	1960	50
1920	70	1970	60
1930	60	1980	70
1940	50	1990	80

* Year of Birth: 1910; confounds age and time of measurement.
** Year of Birth (Cohort); confounds age and cohort.
Note: Cross-sectional design = group "differences"; longitudinal design = group "changes".

of different ages are measured at the same time. The subjects of different ages are born in different years. Thus, they are designated as different "cohorts". In this design, there is confounding of the effects of age with the effects of year of birth, or "cohort". In the longitudinal design, subjects from only one age of birth, or "cohort", are evaluated at successive times of measurement. In this design, time between age, and time of evaluation are confounded (Storandt, 1982). Basically, longitudinal designs attempt to estimate age "changes" within subjects, whereas cross-sectional designs estimate age "differences" among subjects. Table 1 illustrates, by separate examples, the confounding in the cross-sectional and longitudinal designs used for studies of vision and aging.

VISUAL ACUITY, FOVEAL CONE/ROD DENSITY, CELL LOSS IN GENICULOSTRIATE SYSTEM

Because visual acuity plays a fundamental role in visually guided behavior, it has received considerable attention in neuroscience studies of vision, particularly during development. Even a brief review of visual acuity and aging can illustrate the great complexities, opportunities, and challenges confronting visual scientists. Age-related visual decrements, without ocular pathology, are often encountered in clinical settings. The difficulties in interpretation of the onset and magnitude of the effects of age on human acuity declines, based on cross-sectional designs, are generally not fully recognized, in clinical, as well as in research settings. The variations in "normal adult acuity" of the same age, as well as the variability of acuity across different age levels, from maturity to advanced age, have made estimates of optimum acuity, as well as magnitude and rate of subsequent acuity decline, quite difficult to interpret, even in large sample studies (Frisén and Frisén, 1981; Pitts, 1982; see Figure 1). Most current clinically used charts of acuity are based on Snellen's "norm-value" of 20/20, or its equivalents in other notations, developed for assessment of normal adult visual acuity. Age "differences", or less often, "changes" are generally compared with the universally used "norm-value" of 20/20 in the Snellen chart. Many acuity charts do not have letters corresponding to higher acuity levels. There is considerable evidence, however, that normal, young adult acuity, under optimum condition, is considerably better than 20/20 (Frisén and Frisén, 1981). Thus, early detection of age-related acuity declines in cross-sectional designs remains quite difficult because of: 1) the extensive use of the normal acuity norm of 20/20, as optimum acuity, or point of reference, and 2) the great variability of acuity in successive age groups, extending from childhood, to maturity, and to senescence. As a consequence of the extensive use of cross-sectional designs, and the "artificial 20/20 ceiling" on the optimum acuity norm, and the variability of acuity in cross-sectional age groups, clinical and epidemiological studies of acuity and aging have thus far only grossly estimated the onset, magnitude, and rate of acuity decrease in otherwise clinically normal subjects, from maturity to senescence.

It seems reasonable to assume that detection of age-related acuity differences or changes may require a lifespan framework, for studies of vision and aging. Although estimates of lifespan differences, or changes, in human visual acuity should be considered as tentative, they are nevertheless informative as a starting point for studies of vision and aging. Because of the complex interactions of test target characteristics and the optical and neural properties of the eye and geniculostriate system, estimates of visual acuity are quite complex. A number of clinical studies have reported normal acuity of 20/20 to age 60, with a more rapid decline to 80 years of age (Pitts, 1982). Although these estimates of the effects of age on acuity across the lifespan were based on cross-sectional designs, they generally indicate a steep increase in acuity from 10 to 20 years, a relatively stable acuity from 20 to 60, and a more rapid decline from 60 to

80 years of age (Pitts, Figure 8, p. 143, 1982; based on various studies, particularly Weymouth, 1960). It seems relevant to note, however, that comparisons of age and acuity by different investigators also show considerable variability in normal asymptotic acuity, as well as the slope of age-related decline, particularly from 20 to 60 years of age (Pitts, Figure 9, p. 144, 1982). As yet, it remains unclear how variability of acuity changes across the lifespan. Recognizing the complexities of acuity measurements, the effects of the "artificial 20/20 ceiling", as well as the problem of variation in cross-sectional age groups, a more careful attempt has been made to estimate optimum acuity, and the relationship between acuity and aging, using more finely graduated letter charts, with tests conducted under carefully optimized condition, in 100 normal subjects, ranging from 10 to 80 years of age (Frisén and Frisén, 1981). The results of this study indicated a monotonic rise in acuity from a mean of 0.84 at approximately 10 years, to 1.03 by 20, 1.38 at approximately 30, 1.16 around 40, 1.07 by 50, 0.91 by 60, and 0.69 at approximately 70 years. Acuity was tested monocularly, with the natural pupil, using the best correction method. The luminance of the white background was 400 cd/m^2, of the test letters, 25 cd/m^2, with a contrast of 0.88, and room illumination at 300 lx. Least-squares estimates were made of mean acuity in different age groups (Frisén and Frisén, 1981). A comparison of two monotonic acuity functions across 10 to 80 years of age in Figure 1 illustrates graphically optimum adult acuity, and the possible "artificial 20/20 ceiling" effect, which would not be apparent by using many acuity charts that may not have letters or targets corresponding to higher acuity levels. It has been suggested that the use of acuity charts with 1.0 (20/20) as the most acute target may be inappropriate for early detection of acuity decline in the elderly (Frisén and Frisén, 1981). Because of the fundamental importance, and ubiquitous use of acuity in basic and clinical studies of vision and aging, Figure 1 illustrates graphically lifespan differences in human visual acuity plotted as two monotonic functions from 10 to 80 years of age, based on the data cited by Pitts (Figure 8, p. 143, 1982), and by Frisén and Frisén (Table 1, p. 154, 1981).

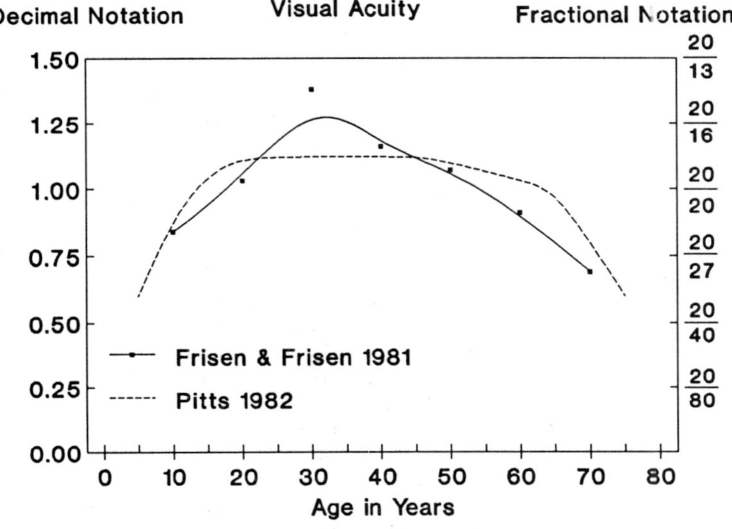

Figure 1. Life span monotonic acuity functions based on cross-sectional age groups from 10 to 80 years.

Classically, it has been assumed that measures of visual acuity involve some type of brightness-contrast sensitivity. In the visual world, however, not all objects are encountered at 100% contrast. Significant age differences in low spatial frequency contrast sensitivity functions (CSF) have been reported between young and older subjects, even though both age groups had comparable acuity with high frequency and high contrast targets (Sekuler et al.,1980). Although visual acuity and aging have been emphasized in this selective review, it also seems important to note that such functions as acuity, color vision, and stereopsis of an individual may not change at the same rate with age. Also, increasing hesitancy, or caution among the elderly in visual testing may result in greater variability, or spuriously low estimates of age-related differences or changes in various visual functions.

A major goal in neuroscience studies of vision during development and aging is based on the expectation that significant correlations can be made between such basic visual functions as acuity and the mosaic formed by foveal cone/rod receptors (Williams, 1986). It appears relevant to note that the terms fovea and macula are commonly used by anatomists and clinicians interchangeably. Also, it is now recognized that only the foveola of the fovea represents a "rod-free" area in the retina of man and nonhuman primates (Youdelis and Hendrickson, 1986). As a starting point for correlating age differences in acuity with foveal cone density, either during development or aging, considerable detail has been presented in this review concerning the apparent wide range, or variability of acuity from an approximate mean value of 0.83, or 20/25, at 10 years, to an estimated optimum value of 1.38, or 20/15, in some individuals by age 30 (see Figure 1). It has been reported that a 2.9-fold range in maximum foveal cone density of young adult human retinae (100,000-324,000 cones/mm^2) may contribute to the wide range of individual differences in acuity (Curcio, et al., 1987, 1990). It has been suggested that this wide range in foveal cone density at maturity may preclude detection of the effects of age on visual acuity and foveal cone density (Curcio, et al., 1990). However, in view of the plausible theoretical considerations (Williams, 1986), as well as the significant correlations among acuity, Nyquist frequency, and foveal eccentricity, it seems unlikely that significant changes in the foveal mosaic would not play an important role in acuity during development, maturity, and aging (Hirsch and Curcio, 1989).

In a preliminary estimate of age differences in cone density of the "rod-free" foveola of the fovea, in the elderly, the number of cones was estimated at 76,282 in a 37 year old, and at 48,804 cones/mm^2, by 72 years, in this central most 250 μm foveola (Youdelis and Hendrickson, 1986). Another study, however, has presented data that there was no significant cone/rod cell loss in the fovea of some subjects from 60 to 80 years (Gao, et al., 1990). It has also been suggested that as yet, there are no quantitative data showing that a significant loss of foveal cones is correlated with a similar loss of acuity, and that rods rather than cones may be more vulnerable to the effects of age (Curcio, et al., 1990). Foveal cone/rod diameter, spacing, and geometry of the discrete sampling "array", however, are generally assumed to determine visual acuity at the level of the retina (Williams, 1988; Hirsch and Curcio, 1989). Foveal cone/rod density in the human retina has been estimated anatomically (Curcio, et al., 1990), as well as by laser interferometric measures in the "living" human eye (Williams, 1988). Because the transduction of the foveal image into neural signals is performed not by a continuous, but a discrete sampling array of cones, their inner segments, and their spacing, problems of psychophysical and anatomical clarification include the existence of visible moiré patterns, or aliasing, when very fine gratings are imaged on the retina. These type of multi-disciplinary problems provide a complex challenge of how foveal cone/rod sampling may set limits of psychophysical acuity resolution, and possibly play a role in trichromacy, or color vision (Williams, 1986). Because of these promising multi-disciplinary developments, more accurate psychophysical estimates of acuity declines in the elderly, and in primate models, will become of special importance as more quantitative morphometric estimates of foveal and foveolar cone density and spacing, and cell number in geniculostriate pathways become available for man (Frisén, 1980; Hirsch and Curcio, 1989; Curcio, et al., 1990) and nonhuman primates (Ordy, et al., 1980, 1982, 1991; Schein, 1988).

The development of appropriate experimental designs remains as considerable challenge for these innovative multi-disciplinary, neurobehavioral studies in which attempts are made to correlate age-related "differences" or "changes" in such specific visual functions as acuity, color vision, and stereopsis, with selected physiological, biochemical, and morphological alterations in the retina and geniculostriate system of man and primate models (Ordy, et al., 1982, 1991). With appropriate designs and adequate sample size, even large variances in acuity and foveal cone density do not preclude detection of significant age changes or differences in acuity or foveal cone density, or in correlations between these two variables across the lifespan. It seems promising to note that "within-subject" correlations may become possible of visual acuity and foveal cone/rod density even in cross-sectional studies from maturity to advanced age, with the development of "non-invasive", laser interferometric estimates of foveal cone/rod densities in the "living" human and nonhuman primate eye (Williams, 1988). Although these studies appear promising, visual acuity declines in aging are also known to be associated with cataract formation, macular degeneration, open angle glaucoma, vascular disturbances, changes in metabolism, loss of melanin, and increase of lipofuscin in retinal pigment epithelium (RPE), as well as possibly other, as yet unknown molecular changes at the level of the eye (Pitts, 1982). In addition to the age-related changes in RPE pigments, photoreceptors, their outer segments or discs, and retinal ganglion cells, that may be causally associated with declines of acuity, it also seems important to note that an estimated loss of 54% of neurons has been reported from age 20 to 87 years, in the foveal projection area of the striate cortex in man (Devaney and Johnson, 1980). Although considerable knowledge is becoming available concerning the physiological and anatomical bases of visual acuity at the level of the human retina (Hirsch and Curcio, 1989), as well as the nonhuman primate geniculostriate system (Wässle, et al., 1987; Schein, 1988), considerably less is known about the genetic, molecular, and environmental causes of its increase during maturation, or its apparent decline from maturity to senescence (see Figure 1 for estimated monotonic increases and decreases of human acuity across the lifespan). Similarly, studies of visual impairments in such age-related neurodegenerative disorders as Alzheimer's disease have only recently begun to focus on the visual system. In addition to memory impairment and cell loss from the hippocampus, some studies have reported retinal ganglion cell degeneration as a prominent manifestation in some Alzheimer's patients (Blanks, et al., 1989).

EPIDEMIOLOGICAL, OPHTHALMOLOGICAL STUDIES OF VISION AND AGING

According to numerous clinical studies, age-related changes in the crystalline lens are closely associated with changes in such diverse visual functions as acuity and accommodation in presbyopia, color vision, contrast sensitivity, glare sensitivity, and field of vision. A variety of environmental conditions influence changes in the lens with age. Onset of presbyopia varies with geographic location (Carter, 1982; Spector, 1982). Other epidemiological and ophthalmological studies of vision and aging have focused mainly on visual impairments more directly related to such disease entities as cataracts, glaucoma, and more recently to age-related macular degeneration (AMD) (Greenberg and Branch, 1982; Lovie-Kitchin and Bowman, 1985). There is a more extensive epidemiological and ophthalmological literature on the incidence of and recovery from ocular pathologies in younger than older subjects. Reviews of methodological issues concerning the incidence and prevalence of visual impairments and ocular pathologies in the elderly have recognized limitations in clinical "case" reports, and other sources of inadequate information or data that may result in "under-reporting" of these visual impairments and ocular pathologies in the elderly (Greenberg and Branch, 1982). Despite methodological sampling limitation, cataracts, glaucoma, and retinal disorders, other than diabetic retinopathy, appear to be six to eight times more prevalent among individuals aged 65 years or more (Greenberg and Branch, 1982). In elderly, low-vision subjects, loss of central vision and loss of peripheral fields of vision appear as major impairments in the context of other "human factors" in aging (Faye, 1982). As yet, there are few comprehensive clinical or epidemiological reviews on how well elderly subjects respond to cataract surgery, or to other forms of treatments of specific ocular pathologies (Lovie-Kitchin and Bowman, 1985).

ENVIRONMENTAL DEMANDS AND HUMAN FACTORS IN VISION AND AGING

Another major issue in vision and aging requiring research, concerns not only the onset, magnitude, and rate of "normal" visual impairments, or diagnoses and treatments of ocular pathologies, but also the individual's and society's expectations, or demands on visual capacities and performance (Faye, 1982). There is for example, a considerable difference, or gap, between an individual's capacity to perform a static visual acuity test in clinical, laboratory, or other test settings, and the capacity to perform such dynamic "real-world" tasks as driving a vehicle (Panek, et al., 1977). Although dynamic, multi-variate tests of visual functions essential for driving are lacking, some studies have shown loss of acuity, contrast sensitivity, decrements in visual field size, visual search performance, and longer recovery from glare in the elderly (Carter, 1982; Pitts, 1982). This example illustrates dramatically that studies of vision and aging appear at an early stage of recognition and development, that they may not be considered in isolation from various environmental demands, and that they may have complex social and policy implications (Greenberg and Branch, 1982).

In the design of optimum environments for the elderly, visual impairments need to be considered in the context of other sensory, cognitive, and motor impairments (Birren and Williams, 1982). There is evidence that social, economic, and physical circumstances may play an important role in delineating the extent to which an individual's visual impairments may be considered acceptable, or disabling (Peterson, et al., 1978). Some novel approaches, such as an "Emphatic Model" (EM) have been proposed as a technique that may "simulate" selected age-related visual impairments while younger observers engage in performance of various everyday environmental tasks (Pastalan, 1982). The rationale for development of such EM models was to "sensitize designers" as to special requirements in visual environments for the elderly, and thus explore novel approaches for making compensatory changes in these environments.

Lighting for the elderly has also been considered not only in terms of vision as a source of spatial information about the environment, but also in terms of photobiological effects, such as its impact on neuroendocrine consequences, vitamin D_3 synthesis, calcium absorption, immunological mechanisms, and cardiovascular regulation (Hughes and Neer, 1981). Designs for lighting for the elderly may need to consider fluorescent light sources that simulate more closely the full spectrum of sunlight with its profound photobiological effects on the health of the elderly.

SUMMARY, CONCLUSIONS

This review was an attempt to provide a brief, and very selective overview of the scope of current scientific, clinical, epidemiological, and social research problems in vision and aging. Basic and clinical scientists studying vision and aging can encounter research challenges and opportunities extending from: 1) molecular events in phototransduction, or conversion of light energy into neural signals in the retina, 2) transformation of an optical image on the retina into neurotransmitters and electrophysiological codes, 3) functional mapping of visual information processing in the geniculostriate system, 4) psychophysical and ophthalmological assessments of visual functions, 5) development of experimental designs appropriate for studies of vision and aging, 6) epidemiological studies concerning the incidence and prevalence of ocular pathology, 7) development of critically needed information on compensatory mechanisms, response strategies, or design of novel visual environments, 8) clarification of environmental demands and human factors in vision and aging, and 9) assessment of photoreceptor damage after long-term exposure of the visual system to light, environmental hazards, toxins, or pollutants, in terms of consequences, and prospects of remedial intervention. It seems clear that visual scientists working in such disciplines as psychology, neuroscience, ophthalmology, neurology, epidemiology, and gerontology can benefit by "crossing" disciplines, thus increasing interdisciplinary communication, and developing better approaches for improving the "visual quality" of life of the increasing proportion of the elderly in the total population.

REFERENCES

Birren, J.E., and Williams, M.V., 1982, A perspective on aging and visual function, in: "Aging and human visual function," R. Sekuler, D. Kline, and K. Dismukes, eds., Alan R. Liss, Inc., New York, NY, pp. 7-19.

Blanks, J.C., Hinton, D.C., Sadun, A.A., and Miller, C.A., 1989, Retinal ganglion cell degeneration in Alzheimer's disease, Brain Res., 501:364-372.

Carter, J.H., 1982, The effects of aging upon selected visual functions: Color vision, glare sensitivity, field of vision, and accommodation, in: "Aging and human visual function," R. Sekuler, D. Kline, and K. Dismukes, eds., Alan R. Liss, Inc., New York, NY, pp. 121-130.

Coleman, P.D., and Buell, S.J., 1985, Regulation of dendritic extent in developing and aging brain, in: "Synaptic Plasticity," C.W. Cotman, ed., The Guilford Press, New York, NY.

Curcio, C.A., Sloan, K.R., Packer, O., Hendrickson, A.E., and Kalina, R.E., 1987, Distribution of cones in human and monkey retina: Individual variability and radial asymmetry, Science, 236:579-582.

Curcio, C.A., Sloan, K.R., Kalina, R.E., and Hendrickson, A.E., 1990, Human photoreceptor topography, J. Comp. Neurol., 292:497-523.

Daw, N.W., Brunken, W.J., and Parkinson, D., 1989, The function of synaptic transmitters in the retina, Ann. Rev. Neurosci., 12:205-225.

Devaney, K.O., and Johnson, H.A., 1980, Neuron loss in the aging visual cortex of man, J. Gerontol., 35:836-841.

Diamond, M.C., and Conner, J.R., 1984, Morphological measurements in aging rat cerebral cortex, in: "Aging and recovery of function in the central nervous system," S.W. Scheff, ed., Plenum Press, New York, NY, pp. 43-56.

Dorey, C.K., Wu, G., Eberstein, D., Farsd, A., and Weiter, J.J., 1989, Cell

loss in the aging retina, Invest. Ophthalmol. Vis. Sci., 30:1691-1699.
Faye, E.E., 1982, The older low-vision patient: Visual function and human factors, in: "Aging and human visual function," R. Sekuler, D. Kline, and K. Dismukes, eds., Alan R. Liss, Inc., New York, NY, pp. 301-314.
Flood, D.G., and Coleman, P.D., 1990, Hippocampal plasticity in normal aging and decreased plasticity in Alzheimer's disease, Prog. Brain Res., 83:435-443.
Foster, D.H., 1990, Inherited and acquired colour vision deficiencies: Fundamental aspects and clinical studies, Vol. 7 of Vision and visual dysfunction, J. Cronly-Dillon, ed., CRC Press, Boca Raton, FL.
Frisén, L., 1980, The neurology of visual acuity, Brain, 103:639-670.
Frisén, L., and Frisén, M., 1981, How good is normal visual acuity? A study of letter acuity thresholds as a function of age, Graefes Arch. Ophthalmol., 215:149-157.
Gao, H., Rayborn, M.E., Myers, K.M., and Hollyfield, J.G., 1990, Differential loss of neurons during aging of human retina, Invest. Ophthalmol. Vis. Sci., 31:357.
Greenberg, D.A., and Branch, L.G., 1982, A review of methodological issues concerning incidence and prevalence data of visual deterioration in elders, in: "Aging and human visual function," R. Sekuler, D. Kline, and K. Dismukes, eds., Alan R. Liss, Inc., New York, NY, pp. 279-298.
Hirsch, J., and Hylton, R., 1984, Quality of the primate photoreceptor lattice and limits of spatial vision, Vision Res., 24:347-355.
Hirsch, J., and Curcio, C.A., 1989, The spatial resolution capacity of the human fovea, Vision Res., 29:1095-1101.
Hubbel, W.L., and Bownds, M.D., 1979, Visual transduction in vertebrate photoreceptors, Ann. Rev. Neurosci., 2:17-34.
Hughes, P.C., and Neer, R.M., 1981, Lighting for the elderly: A psychobiological approach to lighting, Human Factors, 23:65-85.
Kaneko, A., 1979, Physiology of the retina, Ann. Rev. Neurosci., 2:169-191.
Kirschner, C., and Peterson, R., 1979, The latest data on visual disability from NCHS, Vis. Impair. Blind., April:151-153.
Kulikowski, J.J., Murray, I.J., and Walsh, V., 1990, Limits of vision, Vol. 5 of Vision and visual dysfunction, J. Cronly-Dillon, ed., CRC Press, Boca Raton, FL.
Leigh, R.J., 1982, The impoverishment of ocular motility in the elderly, in: "Aging and human visual function," R. Sekuler, D. Kline, and K. Dismukes, eds., Alan R. Liss, Inc., New York, NY, pp. 173-182.
Lovie-Kitchin, J.E., and Bowman, K.J., 1985, "Senile macular degeneration," Butterworth Publishers, Boston.
Lund, R.D., Hankin, M.H., Rao, K., Radel, J.D., and Galli, L., 1988, Developmental and lifespan plasticity, in: "Neural plasticity: A lifespan approach," T.L. Petit and G.O. Ivy, eds., Alan R. Liss, Inc., New York, NY, pp.1-20.
Nathans, J., 1987, Molecular biology of visual pigments, Ann. Rev. Neurosci., 10:163-194.
Ordy, J.M., and Brizzee, K.R., 1979, Functional and structural age differences in the visual system of man and nonhuman primate models, in: Aging, vol. 10, "Sensory systems and communication in the elderly," J.M. Ordy and K.R. Brizzee, eds., Raven Press, New York, NY, pp. 13-50.
Ordy, J.M., Brizzee, K.R., and Hansche, J., 1980, Visual acuity and foveal cone density in the retina of the aged rhesus monkey, Neurobiol. Aging, 1:133-140.
Ordy, J.M., Brizzee, K.R., and Johnson, H.A., 1982, Cellular alterations in visual pathways and the limbic system: Implications for vision and short-term memory, in: "Aging and human visual function," R. Sekuler, D. Kline, and K. Dismukes, eds., Alan R. Liss, Inc., New York, NY, pp. 79-116.

Ordy, J.M., Brizzee, K.R., Wengenack, T.M., and Dunlap, W.P., 1991, Visual acuity, cone, rod, ganglion cell loss, and macular degeneration in the retina of the aged rhesus monkey, in: "The effects of age and environment on vision," D.A. Armstrong, M.F. Marmor, and J.M. Ordy, eds., Plenum Press, New York, NY.

Panek, P.E., Barrett, G.V., Sterns, H.L., and Alexander, R.A., 1977, A review of age changes in perceptual information ability with regard to driving, Exp. Aging Res., 3:387-449.

Pastalan, L.A., 1982, Environmental design and adaptation to the visual environment of the elderly, in: "Aging and human visual function," R. Sekuler, D. Kline, and K. Dismukes, eds., Alan R. Liss, Inc., New York, NY, pp. 323-334.

Peterson, R., Lowman, C., and Kirschner, C., 1978, Visual handicap: Statistical data on a social process, Vis. Impair. Blind., Dec:419-421.

Petit, T.L., and Ivy, G.O., 1988, "Neural plasticity: A lifespan approach," Alan R. Liss, Inc., New York, NY.

Pitts, D.G., 1982, The effects of aging on selected visual functions: Dark adaptation, visual acuity, stereopsis, and brightness contrast, in: "Aging and human visual function," R. Sekuler, D. Kline, and K. Dismukes, eds., Alan R. Liss, Inc., New York, NY, pp. 131-160.

Rodieck, R.W., 1979, Visual pathways, Ann. Rev. Neurosci., 2:193-225.

Scheff, S.W., 1984, "Aging and recovery of function in the central nervous system," Plenum Press, New York, NY.

Schein, S.J., 1988, Anatomy of macaque fovea and spatial densities of neurons in foveal representation, J. Comp. Neurol., 269:479-505.

Sekuler, R., Hutman, L.P., and Owsley, C., 1980, Human aging and vision, Science, 209:1255-1256.

Siegel, J.S., 1980, Recent and prospective trends for the elderly population and some implications for health care, in: "Second conference on the epidemiology of aging," S.G. Haynes and M. Feinleib, eds., U.S. Dept. of Health and Human Services, Washington, D.C., NIH Publication No. 80969.

Spear, P.D., 1985, Neural mechanisms of compensation following neonatal visual cortex damage, in: "Synaptic plasticity," C.W. Cotman, ed., Guilford Press, New York, NY, pp. 111-167.

Spector, A., 1982, Aging of the lens and cataract formation, in: "Aging and human visual function," R. Sekuler, D. Kline, and K. Dismukes, eds., Alan R. Liss, Inc., New York, NY, pp. 27-44.

Storandt, M., 1982, Concepts and methodological issues in the study of aging, in: "Aging and human visual function," R. Sekuler, D. Kline, and K. Dismukes, eds., Alan R. Liss, Inc., New York, NY, pp. 269-278.

Teller, D.Y., and Movshon, J.A., 1986, Visual development, Vision Res., 26:1483-1506.

Van Essen, D.C., 1979, Visual areas of the mammalian cerebral cortex, Ann. Rev. Neurosci., 2:227-263.

Wässle, H., Grünert, U., Röhrenbeck, J., and Boycott, B.B., 1987, Retinal ganglion cell density and cortical magnification factor in the primate, Vision Res., 30:1897-1911.

Weale, R.A., 1982, Senile ocular changes, cell death, and vision, in: "Aging and human visual function," R. Sekuler, D. Kline, and K. Dismukes, eds., Alan R. Liss, Inc., New York, NY, pp. 161-172.

Westheimer, G., 1984, Spatial vision, Ann. Rev. Psychol., 35:201-226.

Weymouth, F.W., 1960, Effect of age on visual acuity, in: "Vision of the aging patient," M.J. Hirsch and R.E. Wick, eds., Chilton Company, Philadelphia, PA, pp. 37-62.

Wiesel, T.N., 1982, Postnatal development of the visual cortex and the influence of environment, Nature, 299:583-591.

Williams, D.R., 1986, Seeing through the photoreceptor mosaic, Trends in Neurosci., 9:193-198.

Williams, D.R., 1988, Topography of the foveal cone mosaic in the living human eye, Vision Res., 28:433-454.
Youdelis, C., and Hendrickson, A., 1986, A qualitative and quantitative analysis of the human fovea during development, Vision Res., 26:847-855.

VISUAL ENVIRONMENTAL IMPAIRMENT SCAN (V.E.I.S.)

William E. Sleight

New England College of Optometry
Boston, MA

INTRODUCTION

Coping with sensory loss is the bane of the elderly. Each day the population over age 65 increases by 1500 people (Kutza, 196_). Many of these individuals will suffer visual acuity loss to between 20/25 and 20/70. This may be due to age-related macular degeneration or because they are waiting for cataract surgery. All too often older persons give up favorite acitivties and necessary tasks because these activities have become difficult to perform.

Visual decline may erode the independence of elderly individuals and may lead to social isolation. Phone numbers may become difficult to recognize, or games may be too taxing to play. Furthermore, threading a needle may become laborious and recreational sewing may be completely abandoned. Difficulty in reading recipes or in pricing foods at the supermarket may lead to over the counter fast foods replacing nutritional meals. Elderly individuals must learn to adapt their daily living strategies in a manner which allows them to continue to remain as independent as possible.

It is a clinical challenge to assess the impact that declining sight has in the daily lives of elderly citizens. The most common measure of vision utilized is Snellen visual acuity. Recently, measures of contrast sensitivity and glare disability have become commercially available. While these measures are useful in determining the visual functioning of the eye as an optical system, they provide relatively little information about how a decline in vision affects an individual's daily visual needs.

Eye care providers are woefully ill prepared to assess the impact that sudden or gradual vision loss has on an elderly person in day to day living. One reason for this is that vision is subjective and each individual responds differently to changes in vision with age. Perhaps the best tool at the clinician's disposal is a thorough case history. Unfortunately, a case history which explores the many aspects of vision in daily living is time consuming and frequently not pursued. The result is many people with mild to moderate visual loss do not receive assistance in coping with visual decline in their lives.

The Effects of Aging and Environment on Vision
Edited by D.A. Armstrong *et al.*, Plenum Press, New York, 1991

Traditionally visual loss has been classified in terms of being legally blind, functionally blind, and partially sighted. A person is said to be "legally blind" if the vision in the better seeing eye, even when fully corrected with ordinary lenses is no better than 20/200 or if the maximum diameter of the visual field is no greater than 20 degrees. A person is said to be "functionally blind" if the vision in the better eye, even with the help of ordinary lenses is no better than 20/70, but is better than light projection or light perception, or if the maximum diameter of the visual field is no greater than 20 degrees. Furthermore, the term "visually impaired" is used when a person is functionally blind or partially sighted and the term "fully sighted" is used for all people who are not visually impaired. By the classification scheme, people with visual acuity reduction between 20/25 and 20/70 are regarded as fully sighted.

The question arises as to what does the term "fully sighted" really mean. Does it mean a person with 20/70 vision is able to perform visually related tasks with as much ease as a person who has 20/30 vision and that the day to day impact of having 20/70 vision is really no greater than the impact of having 20/30 vision. This would be a difficult point of view to defend for two reasons. First, vision is a subjective event and two individuals with a visual acuity decline of 20/30 may not experience the impact of this vision change in the same way. Second, very little information regarding the area of subjective visual impairment with mild visual acuity loss has been published. The result is that the group of people with visual acuity reduction between 20/25 and 20/70 are frequently overlooked in terms of significant visual difficulties. Since they are not recognized at least by definition as being visually impaired, there exists a paucity of information or assistance available to this group.

A gap in the delivery of vision care exists. Efforts should be aimed at identifying individuals with mild visual acuity reduction who are having difficulty with normal daily living tasks. Early rehabilitation must be stressed with introduction to concepts of magnification devices and low vision aids at a much earlier stage than partial sight or legal blindness. This is critical for the elderly who frequently suffer gradual visual loss and gradually disengage from favorite activities with visual demands. By the time they are recognized as having true visual problems they have given up the activities which would provide incentive for wanting to learn to use low vision devices.

This paper explores the relationship between visual acuity reduction of 20/25 to 20/70 and a subjective rating of visual environmental impairment. Several aspects of independent daily living activities have been considered, and a self-rating questionnaire has been developed which reflects the level of perceived visual impairment experienced by elderly citizens. This self-rating scan may be useful for determining which individuals are coping well with their measurable vision loss and for determining which individuals have unsatisfactory visual coping strategies.

DESCRIPTION

An essential element of this study was the development of 97 task descriptions which were compiled into the self-rating questionnaire. These task descriptions were selected to reflect common daily living skills, glare, and driving.

Individual task descriptions were developed in part by a panel of optometrists with experience in geriatric care, and in part were drawn from the results of the study "Visual Environmental Adaption Problems in a Partially Sighted Populations," (S.H. Berry and S.M. Genesky, 1979). The questionnaire format was such that each task description was rated on a scale of 1 through 4 as follows: (1) easy, (2) somewhat difficult, (3) very difficult, (4) impossible.

Visual acuities were measured with Snellen optotypes via projector charts or illuminated acuity charts. Subjects wore their habitual corrections. Only records of those who showed no improvement in acuity with pinhole testing were included in the analysis since such individuals presumably could not be helped by a new lens prescription. Near visual acuities were utilized whenever a task description involved a near point task.

Visual acuity was correlated with the impairment rating for the 97 task descriptions (variables). Correlational analysis was used to examine the data. Fourteen of the task descriptions were selected on the basis of their relevance to common daily living activities and their relative correlation with visual acuity to comprise the final questionnaire.

Our objective was to determine which task descriptions were rated as easy when acuity was not impaired, and which were rated as difficult when acuity was poor. The questionnaire was not of forced choice format. If a subject felt that a task description was not applicable he left it unrated. From the responses received we were able to discover which task descriptions would be broadly responded to by 80% or more of the subjects. Approximately 40% of our population responded to the task ratings in the driving category. Consequently, in this category the p values may have been artificially inflated.

In our initial analysis, three different measures of visual acuity were compared to the impairment ratings. These were acuity of the poorer eye, acuity of the better eye, and the acuity of both eyes seeing together. It is interesting to note that the impairment scores when using the acuity of the poorer seeing eye did not correlate well with the visual acuities. This is presumably because subjects were able to make use of the vision in the better seeing eye and did not perceive the same level of visual impairment as subjects who suffered decreased acuity in both eyes. The correlational results were similar when the acuity of the better eye or the acuity of both eyes seeing together were used. The acuity of both eyes seeing together was used in the final analysis since this was the habitual mode of seeing for all subjects.

RESULTS

The mean age of the test population, consisting of 51 subjects, was 71.3 +/- 6.4 years. Females comprised 58.8% of the group and males comprised 41.2%. None of the test group suffered physical handicaps or mobility impairment. The range of visual acuity scores for both eyes seeing together ranged from 20/15 to 20/70. 90% of the population could read 20/50 or better. The population profile shows the distribution of visual acuities, as illustrated by the histogram in Figure 1.

The analysis revealed that in each category the original 97 task descriptions correlated highly with each other and in essence were providing similar information. In order to avoid redundance, 14 task descriptions were chosen based on their level of significance and their correlational coefficients to comprise the final visual environmental impairment scan. The final task descriptions are presented by category in Table I. The p and r values are given for comparison. These task descriptions were analyzed

Table 1. Task descriptions chosen to comprise the V.E.I.S. Questionnaire

CATEGORY	TASK DESCRIPTION	r VALUE	p VALUE	% RESPONSE
GOING OUT	Recognizing faces across the street on a sunny day	.5180	.000	94
	Reading signs at bus stop	.4895	.000	98
READING	Reading newspaper in bright light	.3883	.008	90
	Reading cooking directions on food boxes or can labels	.3361	.021	92
EATING OUT	Reading the menu posted on the wall in a fast food place	.4876	.001	90
	Following the hostess to the table in a dimly lit restaurant	.3405	.022	90
DAILY LIVING	Threading a sewing needle by hand	.3379	.007	98
	Seeing the tuning dial on the radio	.3851	.006	96
	Seeing minor spots on clothing	.3521	.013	96
	Using a ruler or tape measure	.3054	.014	90
GLARE	Going from a dark room into bright sunlight	.4765	.001	96
	Going by large glass windows	.3184	.042	80
DRIVING	Driving an automobile at dusk	.4117	.005	45
	Seeing road signs while driving	.3570	.087	47

for response differences due to age or sex and no statistically significant differences were found. In other words, an 80 year old female with 20/30 Snellen acuity did not rate her visual impairment significantly different than a 60 year old male with 20/30 acuity.

DISCUSSION

The goal of this study was to develop a self-rating questionnaire which would be sensitive to the relationship between mild visual acuity loss, as measured by Snellen letters, and perceived visual impairment. The results of the correlational analysis show that there is a significant relationship between decreasing visual acuity and increasing perceived visual impairment. However, from a clinical perspective, none of the task descriptions showed a correlational coefficient of r= .70 or greater, suggesting that visual acuity may not be the best predictor of perceived visual impairment.

The relationship of visual acuity to perceived visual impairment is illustrated in Figure 2. The grid shows four possible outcomes for people rating their functional vision. The upper left corner shows people with

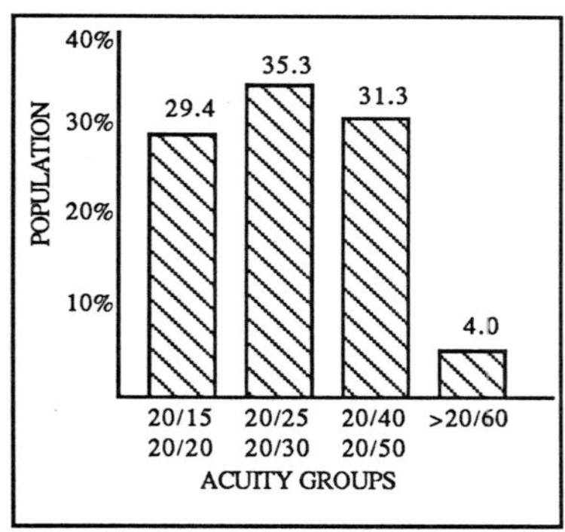

Fig. 1. Distribution of Visual Acuity

SUBJECTIVE CORRELATE

OBJECTIVE CORRELATE		
	GOOD VISUAL ACUITY GOOD SUBJECTIVE RESPONSE	GOOD VISUAL ACUITY POOR SUBJECTIVE RESPONSE
	POOR VISUAL ACUITY GOOD SUBJECTIVE RESPONSE	POOR VISUAL ACUITY POOR SUBJECTIVE RESPONSE

Fig. 2. The relationship of visual acuity to perceived visual impairment

good vision who feel they have little subjective impairment. The lower left corner shows people who have reduced vision but do not rate themselves as impaired. The upper right corner depicts people with good visual acuity who feel that they have significant impairment. The lower right corner depicts people who have reduced visual acuity and rate themselves as significantly impaired. The people represented in the lower left and upper right corners are somewhat paradoxical because their level of visual acuity does not correspond to their level of perceived visual impairment. It may be useful to identify these individuals in order to investigate why this occurs. Perhaps those who do not consider themselves impaired despite reduced vision have given up activities which require fine vision and thus do not feel impaired, or perhaps they have developed alternative methods to continue performing favorite activities. On the other hand, it may be that depression or other psychosocial issues warrant investigation of those people with good visual acuity who consider themselves impaired. The people may also have other visual difficulties such as visual field loss, glare disability, or poor contrast sensitivity.

It must be taken into consideration that self-perception of impaired vision is an individual judgement which may be influenced by non-visual factors. A person's life experiences and personal characteristics may well be important determinants in perceived visual impairment. People with well developed coping strategies or people who are very stoic may not consider themselves impaired to the degree that people who are depressed or who are infirm may be inclined to feel. Visual environmental factors may also play a role in perceived visual impairment. Examples of these factors are poor color contrast, poor indoor lighting, glare from internal reflecting surfaces or dirty windows. Because of these considerations visual acuity may not be the most sensitive indicator of subjective visual functioning. A combined paradigm of visual acuity, contrast sensitivity, and glare disability could well be a better indicator of subjective visual functioning.

CONCLUSION

A self-rating visual impairment questionnaire is proposed. The questionnaire is designed to be a subjective determinant of the quality of visual function. Fourteen indicators of subjective visual impairment were chosen from a field of 97 variables based on their relevance to daily living. When compared to visual acuity, the questionnaire may be useful in identifying elderly people with good or poor visual coping strategies. Future investigation in this area will compare contrast sensitivity and visual acuity to subjective visual environmental impairment using the proposed V.E.I.S. questionnaire format.

REFERENCES

Genesky, S.M., Berry, S.H., 1979, Visual Environmental Impairment Problems of the Partially Sighted, prepared for the Department of Health, Education, and Welfare (CPS-HECS-100) Santa Monica Hospital Medical Center and Rand Corporation.

Kutza, E.A., 1961, The Benefits of Old Age, University of Chicago Press, Chicago, pp. 70-71.

ACKNOWLEDGEMENTS

Consultant in Statistics
Ralph D'Agostino, Ph.D., Boston University
Review Panel
Larry Clausen, O.D., M.P.H.
Alan Lebrecque, Ph.D., O.D.
Karl Streinlein, Ph.D, O.D.

APPENDIX

I. Going Out Category

A. Close Competitors
1. Finding the correct office on a wall directory in an unfamiliar building r=.3931 p=.008
2. Coping with elevator floor buttons r=.3656 p=.001
3. Reading name and number of approaching buses r=.4196 p=.003
4. Determining gate number and departure time from information displays at airports r=.3767 p=.017

B. Weak Relationships
1. Seeing wires or branches at head level
2. Seeing small objects in the path of travel such as skates, toys
3. Seeing sliding glass doors
4. Seeing sprinklers which are mounted in the ground

II. Reading Category

A. Close Competitors
1. Looking up phone numbers in bright light r=.3472 p=.017
2. Reading TV Guide or the TV schedule r=.2994 p=.043
3. Reading recipes from cookbooks r=.4136 p=.005

B. Weak Relationships
1. Reading handwriting in bright light
2. Reading newspaper column headlines in bright light
3. Reading paperback novels in bright light

III. Eating Out

A. Close Competitors
1. Finding out what items cost in a cafeteria r=.4193 p=.005

B. Weak Relationships
1. Reading the check in a dimly lit restaurant
2. Cutting food or buttering bread in a dimly lit restaurant

IV. <u>Glare</u>

A. Close Competitors
1. None

B. Weak Relationships

1. Glare from reflecting surfaces inside hone
2. Glare from bright surfaces outside your home
3. Glare from fluorescent lights
4. Going from bright sunlight into a dark room

V. <u>Daily Living</u>

A. Close Competitors
1. Using a ruler or tape measure r=.3054 p=.041
2. Reading price on food labels or boxes in a supermarket r=.3468 p=.016
3. Reading directions on medicine bottles r=.3645 p=.010
4. Seeing minor spots on clothing r=.3521 p=.013

B. Weak Relationships
1. Applying make-up
2. Matching socks or other articles of clothing
3. Reading a wristwatch
4. Adjusting the thermostat

VI. <u>Driving</u>

A. Strong Competitors
1. Reading the gauges in the car r=.4115 p=.046
2. Seeing moving objects out of side vision while driving r=.3050 p=.145

B. Weak Relationships
1. Spotting potholes or road defects while driving
2. Driving in bright light

VII. <u>Mobility</u>

A. Strong Competitors
1. Walking outdoors on a cloudy day r=.2814 p=.048

B. Weak Relationships
1. Walking outdoors at night on a brightly lit street
2. Walking outdoors in bright sunlight on wet surfaces

ENVIRONMENTAL AND LIFESTYLE CORRELATES OF GOOD VISION IN AN ELDERLY SAMPLE

A. M. Prestrude* and L. G. Meyer

Naval Aerospace Medical Research Laboratory
Naval Air Station
Pensacola, FL 32508

The proportion of the elderly (age 60 and above) in the population of the United States increases each year (Kline, Sekuler, and Dismukes, 1982). Even though the process of aging appeals to very few (the alternative is certainly less attractive), most of us can expect to join the ranks of the elderly. Unfortunately, as our years accumulate, we experience a general decline in sensory acuity, strength, vigor, speed, and beauty. In a society that values these attributes, youth is the standard and the signs of aging are to be avoided, delayed, and when possible, camoflaged. "At present, youth is the fashion and age without honor." Thus, Weale (1963) began the preface to his volume "The Aging Eye." References to human age and the aging process are usually pejorative. A brief scan of "Roget's International Thesaurus" (Chapman, 1977) revealed the following quotes. On youth, "To be young is to be one of the immortals" (W. C. Hazlitt). Old age was called, "An incurable disease" by Seneca and described as, "A tyrant which forbids the pleasures of youth on pain of death" by LaRochefoucauld. Furthermore, old age is considered to be synonymous with debility, fatigue, and weakness. In a similar manner, the verb to age means to become extinct, belong to the past, fade, fossilize, molder, obsolesce, outdate, perish, and rust. Aged means decrepit, infirm, and weak.

In spite of the fact that the proportion of the elderly in the population of the United States continues to increase, the prevailing attitude toward aging has not changed. Few adolescents and young adults are convinced that they are mortal. We are forced to confront our mortality and come to terms with it by the steady accumulation of debilities over the years. Some of these reminders of our mortality are merely annoying and can be ignored or compensated quite readily by minor changes in life style. Others are incapacitating. The majority of these debilities however, reduce the efficiency with which we carry out our daily activities. To some extent, these problems can be remedied by diet, exercise, surgery, and various aids such as corrective lenses. Many of these latter problems are sensory deficits to which we all fall prey as we age and which have been documented in detail by Corso (1971), Fozard, et al. (1977), Panek, et al. (1977), Sekuler, Kline, and Dismukes (1982), Weale (1963, 1983).

Of all the various sensory deficits associated with aging the

*On leave from the Dept. of Psychology, Virginia Polytechnic Institute and State University, Blacksburg, VA 24061.

The Effects of Aging and Environment on Vision
Edited by D.A. Armstrong *et al.*, Plenum Press, New York, 1991

reduction of visual functions most clearly results in a general loss of independence and self esteem (Sekuler et al., 1982). The most commonly occuring visual deficits are reductions in accomodative flexibility, ocular motility, and light transmission by the ocular media and changes in visual neural mechanisms. Uncorrected visual impairments can result in the loss of driving privileges, inability to carry out one's occupation, circumscribe one's leisure activities, etc. Indeed, the elderly driver sometimes is a danger to other drivers and pedestrians, witness, recent segments of the television program 60 Minutes. The visual problems of the elderly driver are compounded by the reduced illumination and contrast of night driving situations (Richards, 1966; 1968; 1972).

Whether age related changes occur in the ocular media or in the retina and visual pathways, the most common result is a reduction in some index of visual acuity. The potential effect of a reduction in visual acuity with increasing age has singular import for commercial and military aviators. From the outset, these individuals represent a group, selected on the basis of visual acuity and general good health, which is superior to the population average. In spite of this, many of these individuals require corrective lenses as they age. For example, the majority of a sample of Air Force pilots with glasses first required them between the ages of 31 and 35 (Hoffman and Koehler, 1958). A decline in visual acuity with age was especially evident after the age of 40 in Air France air crew members (Boissin and LaFontaine, 1973). Even after the age of 50, however, their mean acuity was slightly better than the general population norms. In 1964, 15% of U. S. air transport pilots were age 50 or older and by 1974 that proportion would increase to 50% (McFarland, 1967), a sizable group "at risk" for age related visual and health problems. Since 1959, US commercial pilots have been faced with forced grounding when they reach the age of 60 (FAR 121.383, the "Age 60 rule") because of the increased risks due to the higher incidence of visual deficits and cardiac problems which occur in any over 60 group (Institute of Medicine, 1981). Until 1977, the US Navy had a policy of removing pilots from command pilot status at the age of 45. Currently, their flight status can be determined and changed at any time on the basis of standard visual acuity tests. These policies are not unique to the United States. For example, Schur (1973) pointed out that as early as 1914 the Russian Military Council set the maximum age of military pilots at 45. Currently the West German Air Force limits the age of jet pilots to 40. This age was also recommended as the upper limit for Czech fighter aircraft pilots by Tuma (1967) who also suggested 55 as the upper limit at which pilots should be trained for transport aircraft. European countries have generally enforced a retirement age of 50 for commercial pilots (Hubach, 1958). Hubach, however, recommended that the limit be extended to at least 55 provided such pilots met safety and competency requirements by examination. Early in the history of US aviation experts were of the opinion that flying was a young man's game, but were forced to conclude, after some study, that "Age has little relation to flying ability except when presbyopia begins and reaction time becomes slower" (Wurdemann, 1937). In fact, most experts were willing to concede that the experience of the older pilot often more than compensated for losses in visual acuity and reaction time. The other factor which deteriorated with age and which could disqualify a pilot was cardiac health. The problem of sudden heart attacks and cardiac death apparently has been increasing among commercial pilots (Taylor, 1970).

Perhaps we have labored this issue enough. The point that we wish to emphasize in the present chapter is that, in spite of age related health problems, senescence need not be a time of circumscribed activity and disappointed expectations because of serious visual deficits. The attitudes and policies described above are based on the many observations that general health and fitness, and, in particular reference to the present paper, visual acuity, do decline with age in the general population. Today the standards by which visual acuity is determined are letter charts such as the Snellen letters, tumbling Es, Landolt Cs, or the checkerboard patterns of the Armed Forces Vision Tester and the Bausch & Lomb Orthorater. These are all measures of Static Visual Acuity. In recent years, two other measures of visual acuity have been developed which promise to provide more accurate and valid predictors of real world visual discriminations. They are Contrast Sensitivity and Dynamic Visual Acuity. Contrast Sensitivity measures determine the visual system's ability to transfer information about the various spatial frequencies from very coarse grained to very fine grained patterns such as are represented by sine wave gratings of various spatial frequencies (Campbell and Robson, 1968). Dynamic Visual Acuity measures require that an observer resolve the details of a moving target, for example, detect the orientation of the gap in a Landolt C as it moves by at a given velocity. By comparison, Static Visual Acuity uses only a stationary target of high spatial frequency and, consequently, tests only a small portion of the visual discriminations required in the real world (Leibowitz, Post, and Ginsburg, 1982), especially in aviation and other activities which require detecting and discriminating moving objects such as driving and sports. On the other hand, the similarities of the Dynamic Visual Acuity procedure to the visual tasks involved in driving, flying and sports are obvious (Hoffman, Rouse, and Ryan, 1981).

Given the obvious requirements of good health and good vision for flying, and though much less strictly enforced, for driving, and the general reduction in vision and health with age it seems only logical to investigate the relation between health or fitness and vision. This is particularly pertinent for an aging population (Birren and Williams, 1982). Three recent reports from the Naval Aerospace Medical Research Laboratory (Banta and Grissett, 1985; Banta and Monaco, 1984; Monaco and Banta, 1985) have identified that, in Naval aviation personnel, a high level of aerobic fitness attenuated the expected decline in Dymanic Visual Acuity with age as previously reported by Miller and Ludvigh (1962). Their results (Monaco and Banta) are illustrated in Fig. 1. The correlation between fitness and Dynamic Visual Acuity increased as the target velocity increased to 110°/sec. The age range in these three previous studies was 21-43 and used only Dynamic Visual Acuity as the dependent variable. The present study expanded the age range and used three measures of visual

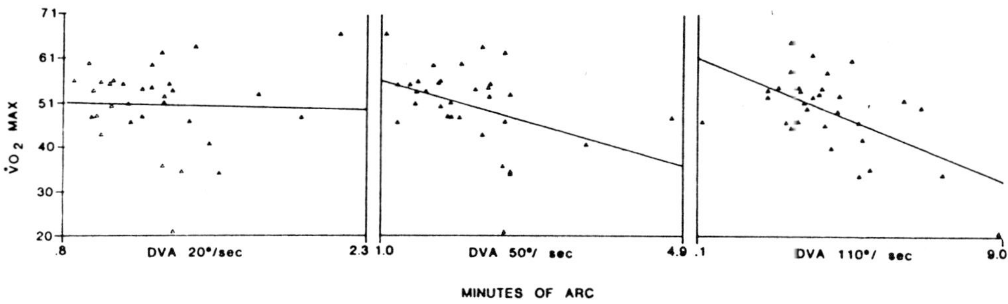

Fig. 1. The relation of Dynamic Visual Acuity to aerobic capacity at three target velocities. From Monaco & Banta (1984).

acuity---Contrast Sensitivity, Dynamic Visual Acuity, and Static Visual Acuity. We compared a sample of retired Naval aviators with two samples of younger individuals.

Our samples were recruited from three sources. Forty Ensigns (mean age = 23 yr, 11 mo) from the Naval Air Station pool, which consists of newly commissioned individuals awaiting assignment, were brought to NAMRL to participate in the vision test procedures. Currently there are approximately 450 retired Naval officers in the Pensacola area. Each of our retired participants was contacted by telephone. Twenty-five of 27 agreed to participate (mean age = 65 yr, 5 mo) and all 25 kept their appointments. Nine of these individuals had previously participated in a study of visual attributes at the NAMRL Laboratory in 1982. Seventeen NAMRL staff members also participated (mean age = 42 yr, 9 mo). The mean ages of our three groups fell into three of the five optimal age groups identified by Braun (1973) for investigating age related variables.

Static and dynamic visual acuity tests were administered with the Automated Dynamic Visual Acuity apparatus recently developed at NAMRL (Molina, 1984). This device projects Landolt C slides (Morris and Goodson, 1983) from a random access projector onto a rotating mirror (Goodson, 1978) which reflects the C target onto a semicircular screen where it is seen moving horizontally in the dynamic visual acuity tests. In the static visual acuity tests, the image of the C target was stationary in the center of the screen. The speed of the mirror, hence the velocity of the target, and the sequence of the target sizes and orientations were controlled by a microprocessor. Thresholds were determined by an up-down or staircase method. In this method each correct response is followed by the next smaller stimulus until there is an error, at which time the next larger stimulus is presented until there is a correct response, etc. The orientation of the gap in the C targets was randomized within sizes. The gap orientation, the response time, whether the response was correct or incorrect, and ten threshold estimates were printed out by the computer at the end of the test which took between 40 and 80 trials. The luminance of the targets was 600 cd/m^2. The contrast level was 33%.

Contrast sensitivity was determined with an Optronix Series 200 Vision Tester. Vertical sine wave gratings of six spatial frequencies were presented in groups of five trials in the order 0.5, 1.0, 3.0, 6.0, 11.4, and 22.8 cycles/degree in a method of adjustment procedure. The screen was illuminated by a Colortran source at 800 cd/m^2 The apparatus determined and printed out an average threshold for each of the six spatial frequencies.

A health evaluation of 26 of the subjects was gathered from a Health Status Questionnaire and 16 of these subjects were also evaluated on a fitness and physiological performance battery consisting of Grip Strength, Percent Body Fat, Sit & Reach (a measure of flexibility), Resting Heart Rate, and Resting Blood Pressure. These measures were included in a test battery recommended by Comfort (1969) to determine physiological aging vs. chronological aging in men.

Overall, we found that our sample of retired Naval aviators, in their 60s, has maintained correctable visual acuity, including dynamic visual acuity, which is comparable to the younger samples, and which is better than previous reports from the general population of 60 year olds. These results are described in the following. Though all the members of our 60 year old sample wore glasses, their static visual acuity was, on the average, only slightly poorer than the standard Snellen acuity of 20/20. It was 1.2 arc minutes which corresponds to a Snellen acuity of

Table 1. Dynamic Visual Acuity by Age Group in arc-minutes

Age Group	Target Velocity	0°	20°	50°	110^{C}	220°
23-11 (n=46)		0.81	1.79	2.85	5.96	12.88
42-9 (n=17)		0.70	1.57	2.24	5.03	15.68
65-5 (n=25)		1.22	2.44	4.67	9.53	26.00
65-0 (n=23)		1.18	2.14	4.28	8.60	23.56
70-5 (n=2)		1.71	2.85	7.85	14.28	36.56

20/24. Furthermore, 11 of the 25 actually exhibited Snellen acuities as good or better than 20/20. A comparison of the static visual acuity of our three samples in terms of arc minutes is presented in the first column of Table 1.

Dynamic visual acuity typically is poorer than static visual acuity and visual acuity declines as the target's velocity increases (Ludvigh and Miller, 1953). There are considerable individual differences in the extent to which target movement causes visual acuity to deteriorate (Miller and Ludvigh, 1962). In the present study, the attenuation of acuity by target movement was the same, on the average, for the 20 year old and 40 year old samples, but individuals in their 60s and 70s were more velocity susceptible, on the average (see Figure 2). These differences were significant at velocities of 110°/sec and 220°/sec. It should be realized however, that very few real world situations involve objects which move through our visual fields at 110°/sec or faster.

Two members of our sample of retired Naval aviators were 70 years of age. When their results were not averaged with the 60 year olds, but considered separately, it can be seen (Figure 2) that the two 70 year olds contributed a large decrement to the average acuity of the group and that the 60 year olds' dynamic visual acuity compares favorably with that of the two younger groups, except at the two fastest target velocities. Recall that nine members of the 60 year old group had been previously tested in 1982. They were tested at target velocities of 20°/sec and 50°/sec (a different apparatus and psychophysical procedure were used). Although the numerical measures were better in the present procedure, there were no significant differences between the average dynamic visual acuity for these nine individuals tested three years apart.

Observers can be divided into velocity susceptible and velocity resistant groups (Miller and Ludvigh, 1962). Velocity resistant 60 year olds exhibited dynamic visual acuity thresholds as good as those of velocity resistant 20 year olds. Velocity susceptible 60 year olds exhibited dynamic visual acuity thresholds as good as those of velocity susceptible 20 year olds, except at 220°/sec (see Figure 3).

Compared to most previous studies of the variation in dynamic visual acuity with age (e.g., Burg, 1966; Heron and Chown, 1967; Reading, 1972), the observers in the 60 year old group of the present study exhibited a smaller decrement in dynamic visual acuity with age (see Figure 4). Direct comparisons among studies are difficult, of course, because procedures vary in illuminance, contrast, target configuration, psychophysical method, and apparatus.

There is an orientation effect in which the down position of the gap in the Landolt C appears to be the most difficult to resolve and the right position is easiest to resolve. These effects become prominent at velocities of 50°/sec and above. The orientation effects are illustrated in Figure 5. There were no significant differences among our three age groups in the relative proportions of Up, Right, Down, and Left orientations resolved at the various target velocities.

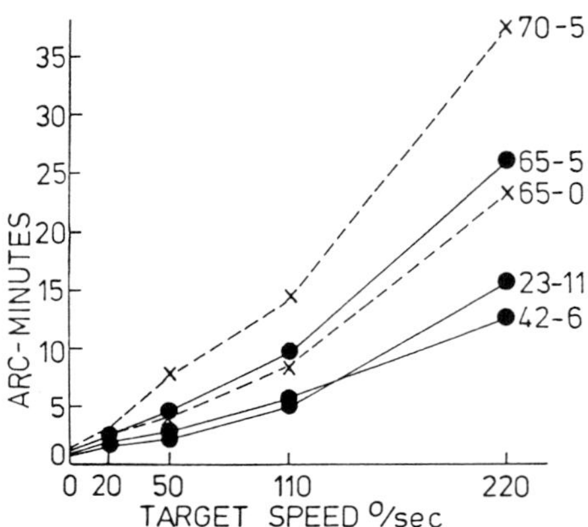

Fig. 2. Static and Dynamic Visual Acuity as a function of target velocity for three age groups. Dashed lines-two 70 year-old observers considered separately from the remainder of the retired Naval aviator sample.

Response times were also analyzed. All groups took longer to respond (185-454 msec, $\bar{X}$=317 msec) on incorrect than on correct responses. The oldest group, on the average, was 78 msec slower on incorrect responses than the youngest group, but there were no significant differences among the three groups in response time for any orientation at any velocity. It should be noted however, that the subjects were not given speed instructions. It is reasonable to expect that our older subjects would have been slower if speed instructions had been given. Indeed, Craik (1969) has recommended signal detection procedures in sensory studies with the elderly because they are allegedly more cautious in reporting their discriminations.

The 60 and 70 year olds exhibited contrast sensitivities that were essentially the same as the youngest group except at the two highest spatial frequencies (see Figure 6). A drop in high spatial frequency resolution is consistent with the observation that our older observers

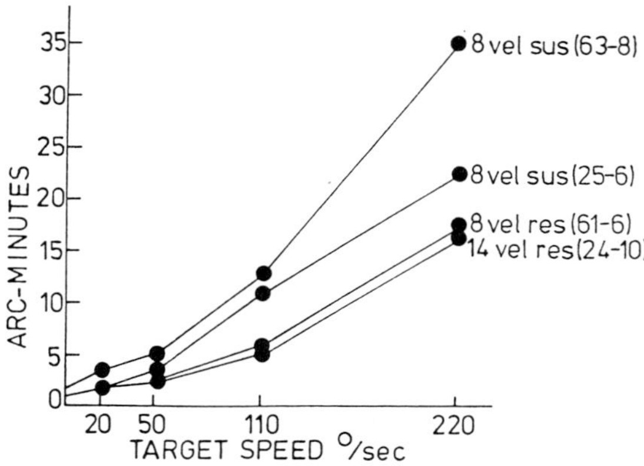

Fig. 3. Dynamic Visual Acuity of the retired Naval aviators of the present study compared with two other studies of DVA in the elderly.

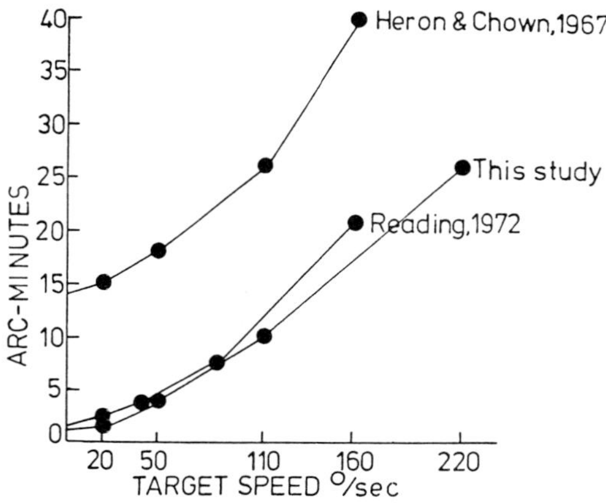

Fig. 4. Dynamic Visual Acuity as a function of target velocity in velocity resistant and velocity susceptible individuals among Ensigns and retired Naval aviators.

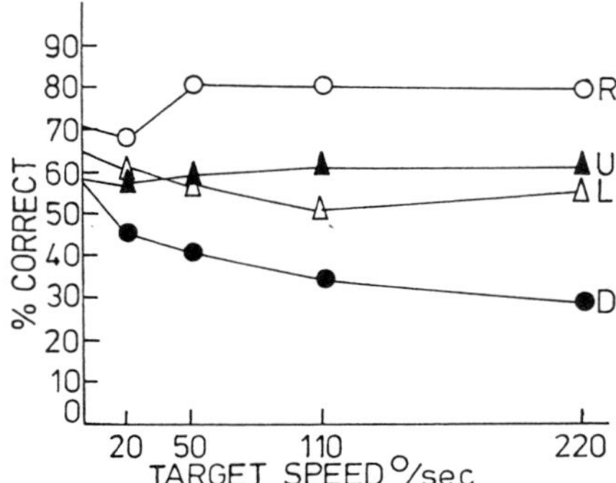

Fig. 5. The orientation effect in Dynamic Visual Acuity measures. Landolt Cs with the gap oriented to the right become easiest to resolve and Landolt C gaps oriented down become most difficult to resolve at target velocities of 50°/sec and faster.

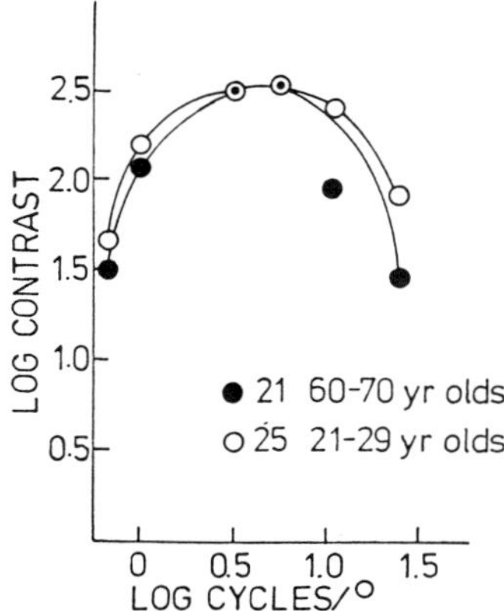

Fig. 6. Contrast sensitivity of Ensigns and retired Naval aviators compared.

Table 2. Health Status Evaluation From Questionnaire and Fitness Measures

	This Study	General Population
Age	48-71	20-44
Height (cm)	177.5	176.8
Weight (kg)	76.3	78.6
Resting Heart Rate	68.6	72.0
Resting Blood Pressure	128/81	120/81
Grip Strength (kg force)	49.1	49.0
Sit & Reach (in)	-2.7	----
Percent Body Fat	19.6	15.0
Smokers	23%	35%

also had slightly poorer static visual acuity compared to the youngest group.

Table 2 presents the health status data from questionnaire and fitness measures taken from 26 of our older subjects (includes one 40 year old and two 50 year olds). Our older subjects are compared to a sample from the general population with an age range of 20-44. Our sample, though a generation older, compares favorably with the younger group.

The 60 year old group from the present study did exhibit a decline in visual acuity, but that decline was not as severe as we have been led to expect from previous studies (Burg, 1966; Farrimond, 1967; Fozard, et al., 1977; Heron and Chown, 1967; Reading, 1972). Three factors could contribute to this observation. First, retired Naval aviators are members of a group that has been selected for extremely good static visual acuity. So, we are, in all likelihood, testing a group with constitutionally good vision. Secondly, these individuals are probably in as good or better physical condition than the average 60 year old male in the general population. Third, their experience as aviators has provided them with training in the kind of target acquisition and resolution skills tested in the dynamic visual acuity procedure.

The latter two factors seem to us to be potentially the most viable explanations for our observations. Recall that recent work from NAMRL (Banta and Grissett, 1985; Banta and Monaco, 1984; Monaco and Banta, 1985) suggested that aerobic fitness attenuates the effect of age on dynamic visual acuity. In the present study 16 of 28 older (40-70) participants were evaluated by both questionnaire and performance tests. Those measures indicated that the 16 individuals were moderately active to sedentary (in comparison to younger groups). The majority had remained active by regularly playing golf (all reported walking rather than riding carts), swimming, walking, bicycling, yard work, calesthenics, etc. On the other hand, three had a history of coronary heart disease, five were hypertensive, and there was one diagnosis each of angina, cancer (in remission), diabetes, and emphysema. However, as a group, they presented the appearance of alertness and vigor. Table 3 presents rank order

Table 3. Rank Order Correlations Between Dynamic Visual Acuities and Fitness Measures (n=19)

Fitness Measure,	Target Velocity	50°	110°	220°
Grip Strength		.52*	.64**	.28
Sit and Reach		.01	-.08	.13
Body Fat		-.44	-.28	-.25
Heart Rate		.05	.02	.15
Systolic BP		.28	.48*	-.10
Diastolic BP		.30	.56*	.36

*p < .05

**p < .01

correlations between Dynamic Visual Acuity and fitness measures from 19 of our oldest subjects. There are significant correlations between the fitness measures of grip strength and systolic and diastolic blood pressure and Dynamic Visual Acuity at 50° and 110°/sec. Grip strength is considered to be an index of vital capacity (Comfort, 1969). Vital capacity has been found to be related to visual field shrinkage with age (Bell, 1972). Blood pressure increases with age (Harrison and Smith, 1979), but there is also a decline in the heart's ability to accelerate blood which probably contributes to the overall decline in vision with age. We obviously need better and more comprehensive evaluations of their cardiac fitness. Due to some administrative problems, we were unable to administer planned maximal exercise stress and pulmonary function tests. These will be carried out with the same group in the future.

Five of the individuals in the velocity resistant 60 year olds and three of the velocity susceptible 60 year olds filled out the Health Status Questionnaire and participated in the fitness measures. All five velocity resistant individuals were above average on Grip Strength and Systolic Blood Pressure. Recall that Grip Strength has been recommended as a measure of vital capacity by Comfort (1969). All three velocity susceptible individuals were below average on these two measures. Grip strength and systolic blood pressure were significantly ($p<.01$) correlated with dynamic visual acuity at 50°, 110°, and 220°/sec.

The parameter of velocity susceptibility merits additional investigation. At this juncture, we cannot attribute status as velocity resistant or susceptible to fitness since our Ensign sample also could be divided into this dichotomy. This suggests the variable of oculomotor motility. That is, accurate tracking of the moving target with a minimum of saccades results in better dynamic visual acuity (Brown, 1972). Of course, better oculomotor motility ought to be one of the consequences of good general health, but there are most certainly other factors involved too. For example, observers do improve dynamic visual acuity with practice (Ludvigh and Miller, 1954). Retired aviators can be considered to have spent 20 to 30 years of almost daily practice in the skills tested in dynamic visual acuity. Perhaps their oculomotor motility is better than that of the average 60 year old which is the reason for their superior Dynamic Visual Acuity. The implications are intriguing in that they suggest that current measures of dynamic visual acuity not only might provide better criteria for the selection and retention of pilots, drivers, etc., but also might be used as a training technique to improve the visual skills of pilots, drivers, and athletes. Samples of Naval aviation cadets (Ludvigh and Miller, 1954; Miller and Ludvigh, 1957) and, more recently, athletes (Burroughs, 1984; Morris, 1977) have been the subject of dynamic visual acuity training. The results have indicated that such training does improve performance on the criterion measure. Whether such training generalizes to other discriminations remains to be demonstrated.

REFERENCES

Banta, G. R. and Grissett, J. D., 1985, April, Relationship of cardiopulmonary fitness of flight performance in tactical aviation. Paper presented at the AGAARO meeting, Athens, Greece.

Banta, G. R. and Monaco, W. A., 1984, Oct., The effects of aerobic fitness on visual performance. Paper presented at the American Aging Association, New York.

Bell, B., 1972, Retinal field shrinkage, age, pulmonary function, and biochemistry, Aging Hum. Devel., 3:103.

Birren, J. E. and Williams, M. V., 1982, A perspective on aging and visual function, in: "Aging and Human Visual Function," R. Sekuler, O. Kline, and K. Dismukes, eds., Alan R. Liss, Inc., New York.

Boissin, J. P. and LaFontaine, E., 1973, Development as regards maximum visual acuity with age among cockpit crew members, Aerospace Med., 44:1075.

Braun, P. H., 1973, Finding optimal age groups for investigating age-related variables, Hum. Devel., 16:293.

Brown, B., 1972, Dynamic visual acuity, eye movements, and peripheral acuity for moving targets, Vision Res., 12:305.

Burg, A., 1966, Visual acuity as measured by dynamic and static tests: A comparative evaluation, J. Appl. Psychol., 50:460.

Burroughs, W. A., 1984, Visual simulation training of baseball batters, Internat. J. Sports Psychol., 15:117.

Campbell, F. W. and Robson, J. G., 1968, Application of Fourier analysis to the visibility of gratings, J. Physiol., 197:551.

Chapman, R. L., 1977, "Roget's International Thesaurus", 4th ed., Thomas Y. Crowell, Publisher, New York.

Comfort, A., 1969, Test battery to measure ageing rate in men, The Lancet, 11:1411.

Corso, J. F., 1971, Sensory processes and age effects in normal adults, J. Gerontol., 26:90.

Craik, F. I. M., 1969, Applications of signal detection theory to studies of ageing, Interdis. Topics Gerontol., 4:147.

Farrimond, T., 1967, Visual and auditory performance variations with age: Some implications, Austral. J. Psychol., 19:193.

Fozard, J. L., Wolf, E., Bell, B., McFarland, R. A., and Podolsky, S., 1977, Visual perception and communication, in: "Handbook of the Psychology of Aging," J. E. Birren and K. W. Schaie, eds., Van Nostrand, New York.

Goodson, J. E., 1978, Dynamics of an image viewed through a rotating mirror, NAMRL Report No. 1252, Naval Aerospace Medical Research Laboratory, Pensacola, FL.

Harrison, W. K., and Smith, J. E., 1979, Age trends in the cardiovascular dynamics of aircrewmen., Aviat., Space and Environ. Med., 0:271.

Heron, A., and Chown, S., 1967, "Age and Function", Churchill, London.

Hoffman, I. L., and Koehler, J. W., 1958, Visual defects in military flyers, J. Aviat. Med., 29:549.
Hoffman, L. G., Rouse, M., and Ryan, J. B., 1981, Dynamic visual acuity: A review, J. Amer. Optom. Assoc., 52:883.
Hubach, J. C., 1958, The old airline pilot, Aeromed. Acta, 5:197.
Institute of Medicine, 1981, Airline pilot age, health, and performance, National Academy Press, Washington, D. C.
Kline, D., Sekuler, R., and Dismukes, K., 1982, Social issues, human needs, and opportunities for research on the effects of age on vision: An overview, in: "Aging and Human Visual Function," R. Sekuler, D. Kline and K. Dismukes, eds. Alan R. Liss, Inc., New York.
Leibowitz, H., Post, R., and Ginsburg, A., 1980, The role of fine detail in visually controlled behavior, Invest. Opthal. & Vis. Sci., 19:846.
Ludvigh, E., and Miller, J., 1953, A study of dynamic visual acuity, NSAM Report No. 562, Naval School of Aviation Medicine, Pensocola, FL.
Ludvigh, E., and Miller, J., 1954, Some effects of training on dynamic visual acuity, NSAM Report No. 567, Naval School of Aviation Medicine, Pensocola, FL.
McFarland, R. A., 1967, Medical fitness for flying in middle-aged pilots, in: "Some Problems of Aviation and Space," Lilienthal, ed., Charles University Press, Prague.
Miller, J. W., and Ludvigh, E. J., 1962, The effects of relative motion on visual acuity, Survey Ophthalmol., 7:83:
Molina, E., 1984, NAMRL automated vision testing devices, in: "Proceedings of the Tri-service Aeromedical Research Panel, Fall Technical Meeting," W. A. Monaco, ed., Naval Aerospace Medical Research Laboratory, Pensocola, FL.
Monaco, W. A., and Banta, G. R., May, 1985, Correlations between aerobic fitness and dynamic visual acuity, Paper presented at the Aerospace Medical Association, San Antonio, TX.
Morris, A., and Goodson, J., 1983, The development of a precision series of Landolt ring acuity slides, NAMRL Report No. 1303, Naval Aerospace Medical Research Laboratory, Pensocola, FL.
Panek, P. E., Barrett, G. V., Sterns, H. L., and Alexander, R. A., 1977, A review of age changes in perceptual information processing ability with regard to driving, Exper. Aging Res., 3:387.
Reading, V. M., 1972, Visual resolution as measured by dynamic and static tests, Pflug. Arch.: Europ. J. Physiol. 333:17.
Richards, O. W., 1966, Vision at levels of night road illumination. XII. Changes of acuity and contrast sensitivity with age, Amer. J. Optom. Arch. Amer. Acad. Optom. 43:313.
Richards, O. W., 1968, Visual needs and possibilities for night driving: Part 12, The Optician, March, 264.
Richards, O. W., 1972, Some seeing problems: Spectacles, color, driving and decline from age and poor lighting, Amer. J. Optom. Arch. Amer. Acad. Optom. 49:539.
Schur, W., 1973, Pilots--middle--age flight fitness, Zeitschr. Militaermed., 10:325.
Sekuler, R., Kline, D., and Dismukes, K., 1982, Aging and visual function of military pilots. Aviat., Space Environ. Med., 53:747.
Taylor, R. F., 1970, Fitness to fly? A case history and brief review of the literature. Canad. Med. Assoc. Jour., 10:745.
Tuma, J., 1967, Several aspects of laboratory psychophysiological activity in the flying personnel of various ages, in: "Some problems of Aviation and Space," Lilienthal, ed., Charles University Press, Prague.
Weale, R. A., 1963, The Aging Eye, H. K. Lewis & Co., Ltd., London.
Weale, R. A., 1983, "A Biography of the Eye," H. K, Lewis & Co., Ltd., London.
Wurdemann, H. V., 1937, Statistics of a thousand physical examinations for aviation, J. Aviat. Med., 8;13.

UPDATE 1990

The aging literature is replete with studies of sensory, perceptual, and cognitive functioning in various elderly samples. Similarly, the salutary effects of exercise and healthful life styles are well-documented. The relation between general health, or more specifically, cardiac fitness, and sensory-perceptual capacity has not been investigated. One study (Colsher & Wallace, 1990) has identified a relation between self-reports of visual and auditory capabilities and cognitive abilities which are largely accounted for by factors such as age, education, and physical health status.

The present author continues to vigorously advocate studying the fitness-sensory capacity relationship because of a major presenting problem. That is the "age 60 rule" currently enforced by both commercial and military aviation. No matter what the health status and sensory capacity, when a pilot reaches the age of 60, that individual can no longer fly as command pilot. This continues to be challenged by the aviation community as being based on inadequate and out-dated research. A similar question is faced by a larger segment of the aging population in regard to driving privileges. These questions can be answered more accurately, and one hopes, more equitably with sufficient information about the relation between fitness and sensory capacity.

REFERENCES

Colsher, O.L. and Wallace, R.B.,1990, Are hearing and visual dysfunction associated with cognitive impairment? A population-based approach. J. Appl. Gerontol., 9, 91-105.

THE EFFECT OF AGING UPON PSYCHOPHYSICAL RECEPTIVE FIELD PROPERTIES

Marcus D. Benedetto and Donald A. Armstrong*

Ophthalmic Research Center, Orlando, Florida and
*Department of Medical Technology, School of Health
Related Professions, Faculty of Health Sciences, UB
Clinical Center, State University of New York, Buffalo,
New York

ABSTRACT

Using the non-invasive techniques established by Enoch (1978) for the examination of psychophysical receptive field properties we have psychophysically examined the effect of aging upon the visual function of the inner retina. Based upon the preliminary work of Johnson, Keltner, and Murr (1980), Johnson, Post, and Keltner (1983), and Armstrong, Benedetto, and Samuelson (1985) there is strong evidence there may be changes in amacrine and horizontal cell function with concomitant changes in age.

INTRODUCTION

The receptive field is one of the basic functional elements in vision. It is defined as that "area of the visual field within which a light stimulus can cause a change in the average firing rate of a single retinal ganglion cell" (Cline, Hofstetter, and Griffin, 1980). Experimentally, receptive fields are usually studied in animals. This is due to the necessity for electrodes to be implanted in the tissue. In humans, this can rarely be done. As a result, human psychophysics must be confined to the examination of receptive field analogs (or psychophysical receptive fields) which are determined from the analysis of tests designed to manifest localized response properties. There are different types of receptive fields as there are different receptive field properties; the focus of this particular research endeavor is the relationship of psychophysical receptive fields to chronological age with respect to retinal eccentricity and surround parameters.

It is well known that the eye undergoes many changes with age (Weale, 1963; Meyer-Schwickerath, 1970; Karpe, Rickenbach, and Thomassen, 1950; Mcfarland, Domey, and Warren, 1950; McFarland et. al., 1960; Kornzweig, Feldstein, and Schneider, 1957). For example, more light is required with increasing age to accomplish the same level of visual task performed at a younger age. Accommodative range is also lost. The structure of the globe changes subtly. With advancing age comes a concomitant diminution of retinal sensitivity and cerebral cortical response. (Weale, 1963) The cornea experiences changes in the endothelium and Descemet's membrane, and changes in shape. The iris becomes more rigid and the pupil shrinks with time. The anterior chamber becomes progressively more shallow which sets the conditions for the onset of glaucoma. The lens keeps increasing in thickness, yellowing, and light scatter. The vitreous undergoes atrophy,

The Effects of Aging and Environment on Vision
Edited by D.A. Armstrong *et al.*, Plenum Press, New York, 1991

leading to detachment and collapse. The amount of macula luteal pigment decreases over time. Peripheral retinal depigmentation of the retinal pigment epithelium can also be noted. This substance functions to reduce light scatter and increase resolving power. However, with age there is a loss of macular acuity. Finally, the orbit also changes, the fat atrophies and the tissue dehydrates yielding a recession of the globe.

The ERG (Electroretinogram) shows a diminution of the b-waves with age (Karpe, Rickenbach, and Thomassen, 1950). Some photopic acuity is lost, mesopic acuity is impaired, and glare tolerance is diminished. Dark adaptation thresholds increase. Color defects not unlike anomalous tritanopia arise. Critical flicker fusion frequency decreases along with visual reaction time and speed of visual performance. In addition to the loss of visual acuity the aged also experience a loss of visual field.

Recently it has been noted that there is a change in visual field with increasing age. In presentations at ARVO by Johnson, Keltner, and Murr (1980) and Johnson, Post, and Keltner (1983) consistent changes in visual field with increases in chronology were demonstrated. More recently, observations by Armstrong, Benedetto, and Samuelson (1985) showed psychophysical receptive field losses in a patient with advanced aging disease. This leads to the notion that there may be changes in receptive fields during normal aging which may account for the visual field changes documented by Johnson, et al.

METHODS

Using the techniques established by Enoch (1978) and applied in Armstrong et al., psychophysical receptive field properties were examined in 90 subjects in various blocks of chronology. These blocks were 5-10, 10-20, 20-30, 30-40, 40-50, 50-60, 60-70, 70-80, and 80-90. Each block consisted of 10 normally sighted individuals with no ocular pathology confirmed by a full ophthalmic examination. Each subject was tested at 0 and 16 degrees.

The testing apparatus (Figure 1) consisted of four sources of light. A circline fluorescent tube provided a uniform background adaptation field, a B & L Autoplot supplied the fixation spot illumination and the mechanism for the location of the area of field to be tested, and one projector provided a small, central test flash presented inside the surround stimulus which was presented by a second projector. This system permitted the examination of visual psychophysical thresholds with respect to the influence of static and dynamic surrounds on the central test flash.

The subject was positioned at the apparatus with his/her chin resting on a chinrest, forehead against a headrest, and hand on the response switch which was connected to a buzzer. The eye to be tested was directed to the fixation spot. The fellow eye was occluded by a light-tight patch. The experimenter set the center test flash such that the subject should see it while looking at the fixation spot. Then the intensity of the surround was increased. The subject was asked to look at the fixation spot and press the switch the instant the center test flash disappeared. The intensity of the surround was increased and then slowly decreased until the central test flash became just detectable. This ascending and descending series was repeated four times for each condition. The conditions consisted of static and dynamic surround with 25/75, 50/50, and 75/25 light dark ratios for each retinal eccentricity - 0 and 16 degrees on the 135 degree meridian. The surrounds consisted of sectored disks of varying sizes calibrated with respect to retinal locus.

Fig. 1. The testing apparatus with stimulus presented on the screen, master control panel on the left, subject is seated with chin in chinrest.

RESULTS

The system of examination used herein is a modification of the Westheimer/Werblin systems used by Enoch (1978). This system explores the summation and inhibition arms of the sustained-like (static) and transient-like (kinetic or dynamic) function. Enoch used these terms to distinguish his psychophysical results from the neurophysiological data which it resembles. However, for the sake of simplicity, in this paper the descriptive terms static and dynamic shall be employed. In most of the psychological receptive field studies, the summation arm has been shown to be extremely robust. Therefore, in order to save time and make testing more palatable to the volunteers, only the more sensitive inhibition arm for both static and dynamic conditions was utilized. This facilitated data collection and presentation. The pattern of presentation herein was similar to that which Enoch employed but has been slightly simplified due to the volume of data to enhance clarity of interpretation.

Figures 2 and 3 show the effect of age upon light/dark ratio for all age brackets (averaged for each bracket) at 0 and 16 degrees, respectively, under the static condition. The x axis is the light/dark ratio and age bracket. In an attempt to minimize clutter on the x axis, 2, 5, and 7 represent 25/75, 50/50, and 75/25 light/dark ratios, respectively. The age bracket is printed directly below each group of 2, 5, and 7. The Y axis is the threshold log candelas per square meter. Across age for both retinal locations, there was a general increase in threshold. For both locations and all age groups, there was a general increase in threshold from 25/75 to 50/50 to 75/25 light/dark ratio. However, with increased age the differences between the light/dark ratios decreased. Although the 25/75, 50/50 and 75/25 light/dark ratio thresholds started at approximately the same level, the peripheral (16 degrees) thresholds were slightly higher than the central (0 degrees) thresholds with advancing age. It is interesting to note that the differences between the 25/75, 50/50, and 75/25 light/dark ratios were slightly greater for the peripheral versus central location across all age groups. This means that the static psychophysical receptive field properties were more affected by age centrally than peripherally. This was despite the fact that the thresholds, with increasing chronology, were slightly higher peripherally than centrally.

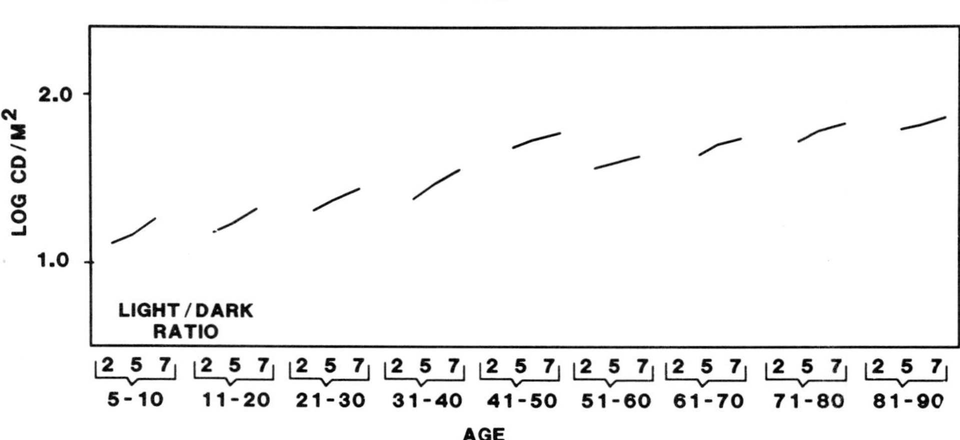

Fig. 2. The results of the effect of age upon static psychophysical receptive field properties at 0 degrees. 2, 5, and 7 represent 25/75, 50/50, and 75/25 light/dark ratios, respectively, for each age bracket. Age bracket is marked underneath. The y axis is threshold in log candalas per square meter.

Figures 4 and 5 show the effect of age upon light/dark ratio for all age brackets (averaged for each bracket) at 0 and 16 degrees, respectively, under dynamic conditions. The x axis indicates the light/dark ratio and age bracket. Here, too, in an attempt to minimize clutter on the x axis, 2, 5, and 7 represent 25/75, 50/50, and 75/25 light/dark ratios, respectively. The age bracket is printed directly below each group of 2, 5, and 7. The y axis is the magnitude of the response, i.e., the difference, in log candelas per square meter, between the static and dynamic conditions for all retinal locations, age groups and light/dark ratios. Across age for both retinal locations, there was a general decrease in magnitude, with a peak between 11 and 30 years of age. For both locations and all age groups, there was a general increase in magnitude from 25/75 to 50/50 to 75/25 light/dark ratios. However, with increased age the differences between the light/dark ratios decreased. Under all conditions, the magnitude of the response peripherally was greater than centrally. Also, the differences between the 25/75, 50/50 and 75/25 light/dark ratios were slightly greater for the peripheral versus central location across all age groups. This means that the dynamic psychophysical receptive field properties were more affected by age centrally than peripherally. Thus, with age, the periphery was shown to be more robust under dynamic conditions.

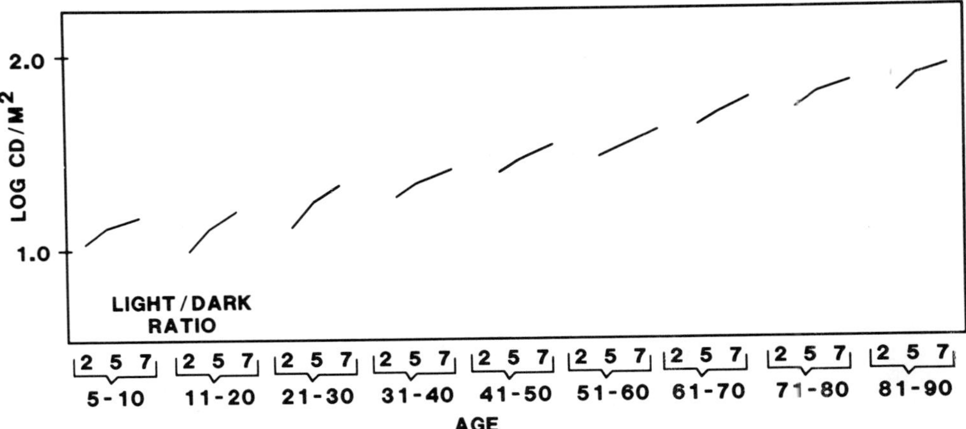

Fiq. 3. The results of the effect of age upon static psychophysical receptive field properties at 16 degrees. 2, 5, and 7 represent 25/75, 50/50, and 75/25 light/dark ratios, respectively, for each age bracket. Age bracket is marked underneath. The y axis is threshold in log candalas per square meter.

Figure 6 is a summary overview of the data comparing the two extreme age groups under all conditions. The x axis is the 25/75, 50/50, and 75/25 light/dark ratios for each retinal location for the 5-10 and the 81-90 age brackets. The y axis is log candelas per square meter. The solid and dotted lines, represent the static and dynamic conditions, respectively. Here, it is particularly easy to see: 1) The magnitude of the response under dynamic conditions was greater peripherally than centrally for both age groups. 2) The magnitude of the response was decreased with age. 3) The difference in light/dark ratio was greater with youth under dynamic conditions. 4) The difference in light/dark ratio was greater in the periphery for both extreme age groups under dynamic conditions. 5) For all static conditions there was a general increase in threshold with an increase in the light/dark ratio. 6) The difference between the static light/dark ratio thresholds decreased with age. 7) Peripheral static thresholds were slightly higher than central thresholds with advancing age. 8) Static light/dark difference thresholds were slightly greater peripherally than centrally.

It is easily noted that the changes from one age bracket to the next are subtle (Figures 2-5). By juxtaposing the extremes, as in Figure 6, observing these differences is facilitated.

DISCUSSION

Until recently, the effects of aging upon psychophysical receptive field properties had not been examined. It had been assumed that normal receptive fields remained unchanged with age as long as the eye was free of pathology. Moreover, it had been previously assumed that receptive fields were similar for all normal individuals as long as blur was carefully corrected.

It is now known that a myriad of visual changes occur with age, such as decreases in sensitivity, acuity, motion perception, visual field, macular function, cortical function, and alacrity of response. All of these changes suggest that the mechanisms underlying visual performance are also changing and are not as stable as previously assumed.

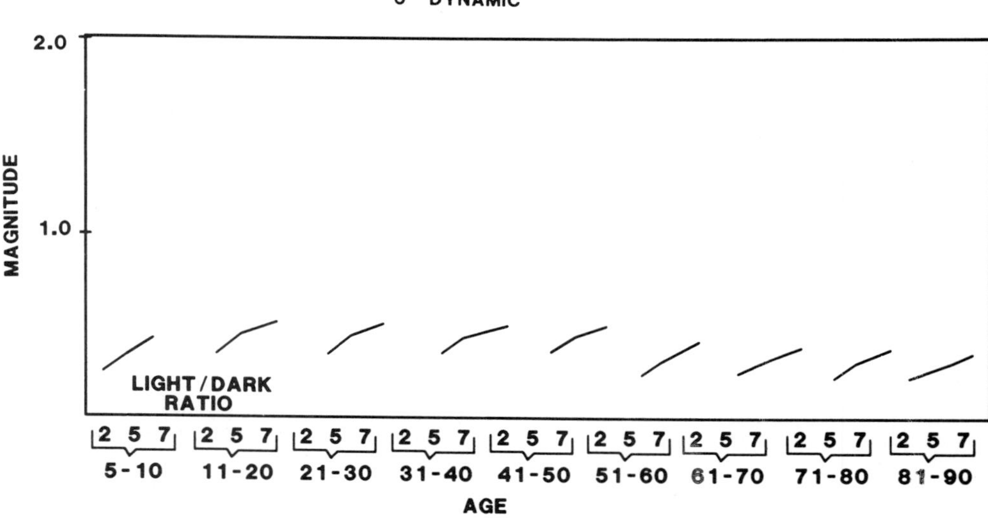

Fig. 4. The results of the effect of age upon dynamic psychophysical receptive field properties at 0 degrees. 2, 5, and 7 represent 25/75, 50/50, and 75/25 light/dark ratios respectively, for each age bracket. Age bracket is marked underneath. The y axis represents the magnitude of the response which is the difference between the static and dynamic conditions in log candelas per square meter.

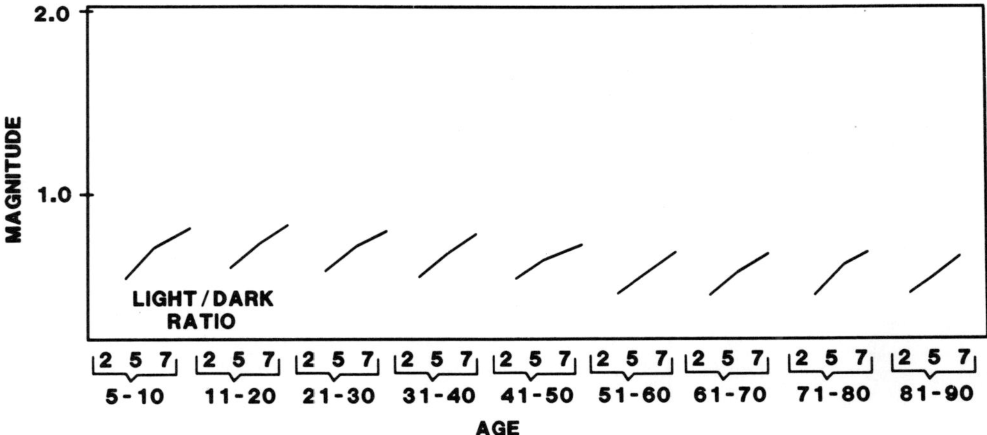

Fig. 5. The results of the effect of age upon dynamic psychophysical receptive field properties at 16 degrees. 2, 5, and 7 represent 25/75, 50/50, and 75/25 light/dark ratios, respectively, for each age bracket. Age bracket is marked underneath. The y axis represents the magnitude of the response which is the difference between the static and dynamic conditions in log candelas per square meter.

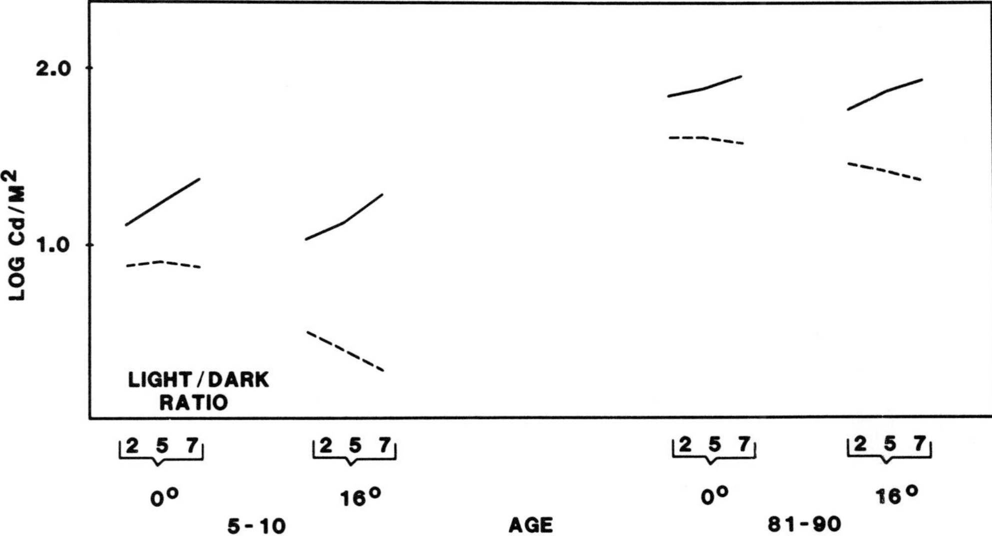

Fig. 6. An overview of the results, juxtaposing the two extreme age brackets for visual comparison. The results of the effect of age upon dynamic psychophysical receptive field properties at 0 and 16 degrees. 2, 5, and 7 represent 25/75, 50/50, and 75/25 light/dark ratios, respectively, for each age bracket. Age bracket is marked underneath. The y axis is threshold in log candalas per square meter. The solid and broken lines indicate static and dynamic conditions, respectively.

This basic information further suggests that changes in the receptive field properties occur as a function of age. This leads to the suspicion that there might be changes in the amacrine and horizontal cell systems of lateral inhibition. If this were true, then these vital systems would not be as responsive and the effective area of neural interaction would change with age, thus altering the receptive fields and their properties.

The recent studies of Johnson, et al. and Armstrong, et al. suggest there may be dynamic processes occurring within a specific layer of the retina, which together with changes in other cell types may account for the changes in visual fields with age. These results, with respect to the Armstrong, et al. and Johnson, et al. data show psychophysical changes with respect to chronology. The changes in visual field and psychophysical receptive field properties, reinforced by the results of this study, suggest there are concomitant neurophysiological receptive field changes which reflect changes in amacrine and horizontal cell function where morphological changes may be evident, i.e., the accumulation of age-pigment. Hence, it might be expected that as age increased, psychophysical receptive field properties would decrease, neurophysiological receptive field properties would decrease, amacrine and horizontal cell function would decrease, and lipofuscin content would increase.

From the data presented herein some inferences may be made with regard to contrast sensitivity. If it were considered that the 50/50 light/dark ratio was representative of a given duty cycle and spatial frequency, then for each retinal location, age bracket, and motion condition, static and dynamic contrast sensitivity trends may be derived. Examining the static 50/50 light/dark ratio condition across age, it may be inferred that, for both retinal locations tested, there is a relatively steady increase in threshold, i.e., a decrease in contrast sensitivity with a concomitant increase in age. The rate of loss of static contrast sensitivity appears to be slightly greater for the peripheral as opposed to the central field location. On the other hand, the dynamic 50/50 light/dark ratio condition across age, for both retinal locations, reached a peak between 11 and 30 years of age, and then began to taper off. The apparent loss of dynamic contrast sensitivity with increasing age was slightly greater for the central visual field location. Whereas, in the more peripheral location, the inferred dynamic contrast sensitivity was more robust, i.e., less affected by age. This difference between central and peripheral changes with age was more dramatically shown in Figure 6 which represented a cursory overview of the data by the juxtaposition of the youngest and the oldest age brackets from all conditions.

A practical clinical implication from this data, based upon the fact that with age the periphery remains robust for motion, would be to try to exploit this aspect. Patients could be trained to utilize their central vision for static tasks and their peripheral vision for dynamic visual tasks. Or to invoke both systems to optimize visual performance. One way to accomplish this would be to convert the nature of a given visual task to utilize the robust dynamic system. An extreme example would be reading which is normally a static task utilizing central vision. It may be discovered that a slight movement of the material will facilitate the task. (As a caveat, we have found this to be successful in working with the visually impaired population.)

The results of this study indicated that the static and dynamic psychophysical receptive field properties, in particular the inhibition arm (static and dynamic), was sensitive to changes in age. At a later date the differential changes between the dynamic and static, centrally and peripherally, may prove to be diagnostic or even prognostic with respect to determining normal or abnormal aging. This may permit the clinician to determine if the patient is aging appropriately or to delineate the degree to which the difficulty which a patient may present is due to the natural aging process and how much is due to other factors or a disease state.

Finally, since this test (psychophysical receptive field properties) is sensitive to changes in age and, especially since it is non-invasive, this functional unit of vision may prove to be a valuable tool as a means to measure physiological versus chronological age. Of course, further study is necessary for confirmation.

ACKNOWLEDGEMENT

This research was supported in part by a grant from the Glenn Foundation for Medical Research.

BIBLIOGRAPHY

Armstrong, D., Benedetto, M., and Samuelson, D., 1985, Amacrine and horizontal cell dysfunction in adult ceroid-lipofuscinosis (Kufs Disease) and anatomical correlates in the ovine model. Int. Ophthalmol., 8:37.

Enoch, J., 1978, Quantitative layer-by-layer perimetry. Invest. Ophthalmol., 17:208.

Johnson, C., Keltner, J., and Murr, W., 1980, Mass visual field screening in 20,000 eyes. ARVO, Orlando, April.

Johnson, C., Post, R., and Keltner, J., 1983, Properties of the normal visual field for different age groups. ARVO, Sarasota, May.

Karpe, G., Rickenbach, K., and Thomassen, S., 1950, The Clinical ERG: 1, The normal ERG above 50 years of age. Acta Ophthal. 28:301.

Weale, R., 1963, The Aging Eye. H.K. Lewis & Company Ltd. London.

Meyer-Schwickerath, G., 1960, Light Coagulation (translated by S.M. Mrance). C.V. Mosby Co. St. Louis.

McFarland, R., Domey, R., and Warren, A., 1960, Night Visibility: dark adaptation as function of age and tinted windshield glass. Highway Res. Board Bull. 255-47.

McFarland, R., Domey, R., and Warren, A., 1960, Dark adaptation as a function of age. J. Geront. 15:149.

Kornzweig, A., Feldstein, M., and Schneider, J., 1957, The eye in old age. Am. J. Opthalmol. 44:29.

AGE-RELATED DIFFERENCES IN ICONIC MEMORY AND VISUAL ATTENTION

David A. Walsh

Department of Psychology
University of Southern California
Los Angeles, California 90089-1061

INTRODUCTION

The purpose of this chapter is to examine age-related differences in iconic memory and visual attention. In order to accomplish these goals, I will first present a brief conceptual and historical introduction to iconic memory and visual attention research. The purpose of this overview is to explicate the nature of these two phenomena and to provide a glimpse at the experimental tasks used to study them. Next, I will review research in aging that has examined iconic memory and discuss the implication of those findings for everyday seeing activities. Finally, research in aging and visual selective attention will be reviewed along with some new findings from my laboratory. The significance of age differences in selective attention will be discussed and it will be concluded that age-related differences in these processes have tremendous significance for everyday seeing activities.

In 1960, George Sperling published his investigations of the information available in brief visual presentations.[23] A major findings was the existence of a brief visual memory that lasts for a few tenths of a second. Neisser labeled this visual storage "iconic memory" and the label has become well accepted.[19] The question Sperling set out to investigate was: How much can be seen in a brief visual exposure? This question has relevance to our normal mode of seeing, which resembles a series of brief exposures as, for example, in reading where the eye assimilates information only in the brief pauses between its quick saccadic movements. Many researchers prior to Sperling had investigated this question, and found that when displays containing a large number of letters were presented to subjects they could report only an average of 4.5 items, although they insisted they had "seen" more. Sperling was able to resolve this paradox with some clever methodology.

The standard methodology used by previous researchers was the "whole report" method: Subjects were shown tachistoscopic displays and asked to report all of the items presented. Sperling's investigations used a "partial report" procedure: Subjects heard tones varying in frequency which signaled a sub-portion of the display to be reported. Three tones were used in combination with displays arranged in three rows of four letters. A high tone indicated the top row was to be reported, a low tone indicated the bottom row and a medium tone signaled the middle row. All tones _followed_ the offset of the displays and signaled subjects to "attend" to a particular sub-set of the display. Sperling's results supported the claims of subjects that they had been able to see all of the display in a single brief glance: Subjects accurately reported the randomly chosen portion of the display they had been instructed to attend to, if that portion was signaled immediately following the display offset. Sperling also examined the accuracy of subjects' partial report performance as the signal tones were delayed. The high accuracy of partial reports gradually declined to the level of whole report performance as the signal tone was delayed from 25 to about 300 msec.

Sperling's findings have two important implications for understanding how people perceive tachistoscopic displays. First, the gradual decline in partial report performance as a function of delaying the signal tone suggests the existence of a brief memory that decays over time. Studies by Sperling, Averbach and Coriell, and others have confirmed the existence of a brief memory for tachistoscopic displays and have shown it to be visual in nature [2,6,12,13] Second, the superiority of "partial reports" to "whole reports" points to some limits in human visual processes. Neisser [19] has provided an insightful analysis of these processing limits that centers on attentional and recognition capabilities. He hypothesized that successful performance of the partial report task requires four separate cognitive elements: 1) iconic memory, the very brief visual memory discovered by Sperling; 2) selective attention, a process that locates the "to-be-recognized" element from among the multiple elements in a display; 3) focal attention, a process that concentrates pattern recognition processes on the "to-be-recognized" element; and 4) pattern recognition, the processes responsible for identifying the target element. This theoretical frame work suggests that the limit of 4.5 items on whole reports may result from the combined limits of iconic memory duration (about 250 msec in length) and the processing time required by pattern recognition, focal, and selective attention processes. Perhaps subjects can report only 4.5 items because the time required by attention and pattern recognition processes exceeds the 250 msec duration of iconic memory--the icon has faded before a fifth or sixth item can be recognized.

AGE-RELATED DIFFERENCES IN ICONIC MEMORY

A substantial body or research now supports the idea that attention and pattern recognition processes take considerable amounts of time, and that these time requirements place a limit on the number of elements that can be reported from multielement tachistoscopic displays.[2,3-10] Averback and Corriell [2] demonstrated that the processing time requirements associated with partial report performance serve to confound the estimates of icon duration obtained with these procedures. This finding is especially relevant to studies of aging and iconic memory. It is well known that older adults require more time than young adults to complete many cognitive processes.[24] To the extent that age is associated with longer attention and pattern recognition processing times, estimates of iconic duration made with partial report procedures will underestimate the duration of an older adults' iconic memory and thereby exaggerate age differences in this measure. The extreme case of confounding has been reported by researchers who found that adults over the age of 70 are unable to perform the partial report task--leading to two possible conclusions--either they have no iconic memories, or their attentional and pattern recognition processes are so slow as to prevent items from being read from iconic memory.[1,20,25]

Fortunately, some researchers [6,12,13] have recognized the limits of the partial report procedure and have developed alternative "direct" measures of iconic persistence that have proved useful in examining age differences in iconic persistence.[15-18,26] Kline and his colleagues employed a visual integration task (the "stimulus halves" task) to examine age differences in iconic persistence.[15-18] Kline and Baffa [15] presented 21 and 56 year old age groups with two successive tachistoscopic stimuli, which formed one of five three-letter words when they were superimposed on each other. The stimulus halves were separated by interstimulus intervals (ISIs) ranging from 0 to 150 msec varying in 30 msec steps. The results of their study showed that the older adults had shorter iconic persistences than the young. The old recognized fewer word trigrams at each ISI than did the young--thus the persistence of the first half appeared to have faded more quickly for the old and was less available to be integrated with the second half, an integration that was necessary to recognize the word formed by the combination.

However, just the opposite result was reported by Kline and Orme-Rogers [16] using a modified version of the "stimulus halves" task. These investigators reported that older subjects did much better than the young in identifying the target words when the two halves were separated by ISIs of 60 and 120 msec, suggesting that the iconic persistence of the first half was longer in duration and/or of higher fidelity for the old. How can these two disparate outcomes be explained? Kline has argued that the major factor is the amount of perceptual closure required by the stimulus displays in the two studies. The study finding a persistence disadvantage for old adults employed displays composed of dot patterns whereas the later study used displays composed of line segments. The dot pattern displays required perceptual closure for successful recognition whereas the line segment displays did not. Thus, Kline's interpretation is based on a logic similar to that outlined for weaknesses in the partial report task--age differences in the cognitive processes required by the task produce a confounded estimate of age differences in iconic persistence.

Another "direct" measurement approach asks observers to make subjective judgements about the visual persistence of a target form that is cycled off and on.[12] The typical procedure involves tachistoscopic presentations of a fixed target duration separated by varying ISIs in a continuous cyclical series. A threshold of persistence is obtained by taking an average of some number of ascending (the ISIs increase from zero) and descending (ISIs decreasing from perhaps 1000 msec) threshold estimates. Walsh and Thomspon [26] reported a consistent and reliable persistence advantage of about 15% for young subjects as compared to old adults across six separate viewing conditions. However, Kline and Schieber [18] reported a persistence advantage of about 25% for the old in a very similar task employing viewing conditions with white contours on a black background and the inverse.

While it is difficult to know what to think of the conflicting pattern of results regarding age-differences in iconic memory,[15-18,26] the potential implications of these findings for everyday seeing seem clear: Age-related differences in iconic persistence are unlikely to explain many of the molar age-differences seen in visual perception. Contemporary models of visual information processing assume that iconic memory performs a buffering function that prolongs the availability of visual information in order for pattern recognition process to complete successful identifications.[19,24] Based on the research reviewed above, it seems safe to conclude that aging is not associated with any large decline in iconic persistence and is, perhaps, associated with an increase in this persistence. Thus, it seems unlikely that the age-related differences found in the studies reviewed here will prove very useful in explaining age-differences in visual perception as conceptualized from the information processing perspective.

AGE-RELATED DIFFERENCES IN VISUAL ATTENTION

Over the last decade researchers have grown to recognize a number of different types of visual attention phenomena. Selective and focal attention as manifest in the partial report task have been described above. A third type of visual attention is involved in the prolonged monitoring of continually changing visual displays. This mode of attending, also known as vigilance, has been the topic of recent research in my laboratory and the outcome of those studies will be reported below. First, however, I will review what is known about age differences in selective and focal attention processes.

Sperling's introduction of the partial report task for the study of tachistoscopic perception of multielement displays stimulated much interest in visual attention.[2,3-10] It seemed surprising to think that an observer could shift their attention among various items in less than a few 100 msecs, much less succeed in identifying the attended to item before an ephemeral icon faded. A program of research by Eriksen and his colleagues [3-10], and others [2,19], mapped the time course and limits of these facile attentional processes. Early investigations of age differences in iconic memory using the partial report procedure [1,20,25] suggested the existence of large age differences in the time course of selective and focal attention processes. An early strategy employed to examine these age differences was to break the partial report task down into its hypothesized attentional and recognition components and to study these components in simple, single-element displays.[24,25] In general, this research suggested that old adults required disproportionately longer processing times than young adults to complete tasks requiring the coordinated utilization of selective attention, focal attention and pattern recognition.

Another line of research examined age differences in the rate at which successive elements (letters or digits) could be recognized or "read-out" of tachistoscopic displays.[24,25] Displays were kept relatively simple (one to three items) to assure that older subjects could perform the task. These studies found that older adults were slower than middle-aged adults, who were in turn slower than young adults in completing pattern recognition processes (estimated processing times of 52, 31, and 24 msec per item for old, middle-aged, and young, respectively). The estimates of pattern recognition processing speed were derived from the slopes of linear functions computed between the number of elements in the displays and the ISIs between the displays and a trailing mask that were required to accurately report the targets. The data of the young and middle-age subjects were fit better by linear functions than were those of the old. The departure from linearity in the old subjects' data resulted from a larger increase in the time to recognize a third as compared to a second letter. A possible explanation for this finding is that older adults may be more sensitive than younger age groups to the distracting influence of adjacent items in multielement displays.[7] This possibility was investigated in a recent investigation in the author's laboratory.

A single-frame visual search task was used to examine the hypothesis that older adults are more sensitive than young adults to the distracting effects of adjacent items in multielement visual displays. Twenty-four young (mean age = 18.6 years) and 24 old adults (mean age = 70.3 years) searched for either a consonant from the alphabet or a digit in displays composed of either 4, 6, 8, or 12 items. The non-target or distractor items in all displays were consonants. On half of all trials a target was present in the display. Whether a display contained or did not contain a target and where the target was located, when present, were determined randomly. All items within a display were located equal distance from one another along the circumference of a circle. This arrangement assures that each display element is equal distance from a central fixation point, both within a particular display and across displays of different size. Thus, any interactions of age with display set size can not be explained simply by age differences in the quality of a peripheral vision gradient, since the retinal location of display items does not change with display set size. Instead, the interaction of age with display set size could be attributed to the amount of distraction provided by adjacent items which are located closer to one another as set size increases from 4 to 12 items.

Each search trial began with the display of a single target (either a consonant or digit depending on the between subjects experimental condition). When the subject pressed a ready button, the target display was replaced by a fixation field one second in duration which, in turn, was followed by a 100 msec presentation of the search display. The subjects' task was to press one of two keys, which signaled whether they had or had not seen the target in the search display. Accuracy of response and reaction times were recorded by a PDP 11/34 computer. The matrix of correct and incorrect detections and rejections was used to compute D' (a signal detection measure of the subjects' sensitivity to the targets, unconfounded by differences in response criterion). The search displays were presented on a Data Media Elite 1521 CRT using a P4 phosphor which was connected to the PDP 11/34.

Figure 1 shows the results of this single-frame search task for the D' measure (the outcomes for reaction time are identical to those for D' and will not be discussed further). There are three statistically significant ($p < .05$) main effects present in these data: 1) The old subjects were less sensitive than the young in detecting targets; 2) detection sensitivity decreased for both age groups as frame set size increased; and 3) detection was better for digits than consonants. None of the possible two way or three way interactions were statistically significant. These data offer no support for the hypothesis that spatially adjacent distractors create more of an attentional processing drain on the performance of old as compared to young subjects. Support for this hypothesis would have been manifested in an age by frame set size interaction, and one would have expected to see an increasing divergence between the detection performance of the young and old as frame size increased. There is little or no evidence for such an effect in Figure 1. Instead, the outcome of this research suggests that aging is associated with a general decline in the resources of attention and detection rather than to an increased sensitivity to distraction from spatially adjacent items.

A second line of recent research has explored age differences in vigilance processes. Shiffrin and Schneider have demonstrated two qualitatively different types of vigilance performance.[21,22] The first, controlled search, demands a constant effortful allocation of attention and appears in conditions in which the to-be-detected targets are intermixed across trials with distractor elements of a display (a procedure called varied mapping). The second, automatic detection, is an effortless processes that is relatively unaffected by task load factors such as the rate of presentation or the search display size. In contrast to controlled search, automatic detection performance depends on a consistent mapping of target and distractor items--targets are selected from one subset of materials while distractors are selected from a different subset. Furthermore, automatic detection emerges from a developmental process that requires large amounts of practice. The developmental process involves a gradual transition from controlled search (where task load factors severely impact detection performance) to the automatic detection state (where task load factors are less important).

The vigilance component of the Shiffrin and Schneider research is introduced by transforming a single frame search task into a multiple frame task. In their work, and in the present research, subjects searched for targets through 20 successive display frames on each trial. In our research a CRT was used to present the 20 successive search frames composed of four letters each. Subjects looked for either of two targets selected from the first nine consonants of the alphabet in displays composed of distractors selected from the last nine consonants. The target letters for a particular trial were selected randomly from the target set and displayed at the beginning of the task for a variable period controlled by the subject. The subjects pushed a button when they were ready to begin a trial. The button press erased the two targets, presented a central fixation point for 1.5 seconds which was followed in turn by 20 successive display frames (the central fixation point remained present while the display frames changed around it). Subjects pressed a "yes" button as soon as they detected a target and a "no" button at the end of the 20th frame if neither of the two targets had been detected. The subjects' reaction times for detection were recorded as well as the matrix of correct and incorrect detections and rejections. This matrix was used to compute D', a measure of the subjects' sensitivity at signal detection free from response biases.

The subjects in this research were 15 dedicated adults who worked at our vigilance task for one and a half hours a day on 18 successive business days. Five of the subjects varied in age from 22-28 years (average age = 24), another five from 41-43 (average age = 42), and the remaining five from 59-68 (average age = 64). On each of the 18 days subjects performed in 300 separate 20 frame trials (each subject thus viewed 5400 individual trials for a total of 108,000 display frames). The 300 trials of each day's performance were divided into five blocks of 60 trials. The first three blocks of each day were presented to subjects with a 200 msec display frame duration--the first frame was visible on the CRT for 200 msecs, it was followed by the second frame without an interstimulus interval, which also remained visible for 200 msecs. Likewise, the third frame followed the second, etc, until the twentieth was presented--the screen was cleared after the last frame had been displayed for 200

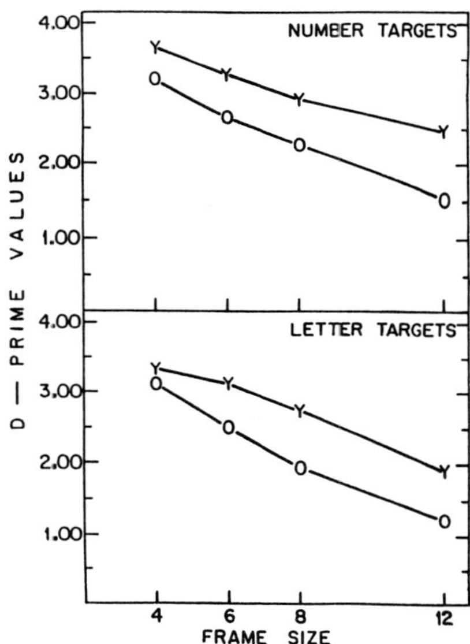

Fig. 1. Target detection sensitivity for Young (Y) and old (O) adults in a single frame visual search task. Subjects searched for either a single digit or letter among letter distractors in displays containing 4, 6, 8, or 12 items.

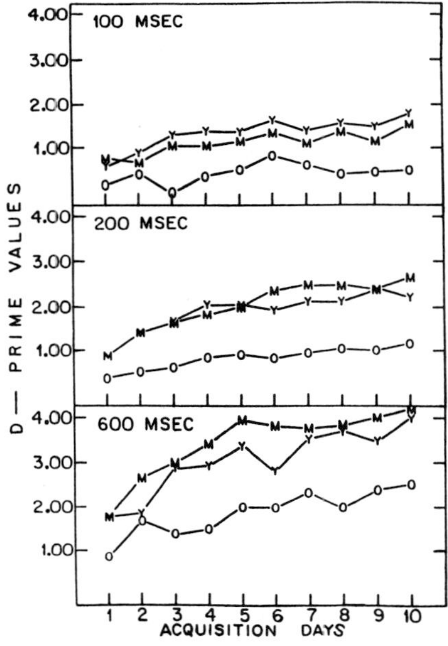

Fig. 2. Target detection sensitivity for young (Y), middle-aged (M) and old (O) subjects in a 20-frame visual search task using consistent mapping of target and distractors. Performance is shown across 10 days of practice for three display durations (100, 200, and 600 msecs).

msecs. The other two 60-trial blocks differed only in the frame duration used: One block used a 100 msec duration, the other a 600 msec duration.

The purpose of the 100 and 600 msec duration trials was to measure the effects of task load (as manipulated by time constraints) and thereby assess the type of vigilance process a particular subject was operating with. (As pointed out above, automatic detection processes are relatively insensitive to task load factors whereas controlled search is very sensitive to these factors.)

One final manipulation remains to be described before discussing our findings. Recall that a consistent mapping procedure using the first nine consonants as targets and the last nine as distractors was employed. On the eleventh day we switched the target and distractor sets, so that distractors were now selected from the first half of the alphabet and targets were selected from the second half. This reversal of target and distractor sets was included to provided a second assessment (in addition to task load) of the degree to which different age groups had developed automatic detection of the target items across the first 3,000 experimental trials. Evidence for age-related differences in the development of automatic detection would be expected to show up in an age by acquisition/reversal interaction, with the groups developing the most effective automatic detection during acquisition showing the greatest impairment in early reversal trials.

Figure 2 shows the target detection sensitivity (D-Prime) data for each of the three age groups across the first ten sessions of the experiment. The middle panel shows performance for the 200 msec frame duration trials (each point is based on three 60-trial blocks), the upper panel shows the 100 msec frame duration trials and the lower panel shows the 600 msec frame duration trials (each point in these two conditions is based on one 60-trial block). In general, the young and middle aged groups performed similarly across all conditions of the experiment. These groups outperformed the old group both in first day detection and in the rate of detection improvement across acquisition days. However, there is little evidence for any age differences in the development of automatic detection responses in Figure 2. We expected to find an age X display rate X acquisition day interaction--an interaction that would have shown that the old gained and lost more in detection performance on the last as opposed to the first acquisition day than the younger groups from increasing and decreasing the frame duration to 600 and 100 msec, respectively. There is no evidence for such an effect, on the contrary, the old subjects show losses and gains in performance that are virtually identical to those of the younger groups. Also, the proportionality of the loss and gain in detection performance is about equal on the first, fifth and tenth days of acquisition.

Thus, the data of Figure 2 point to large and stable age differences in vigilance performance and suggest that these differences increase with extended practice (3000 trials in the present experiment). However, the data of Figure 2 do not support the idea that old adults develop automatic detection responses more slowly than the young. The primary evidence for automatic detection in the performance of a subject is the insensitivity to task load factors. We expected to find less gain and impairment in performance resulting from increases and decreases in frame duration to 600 and 100 msec as acquisition days advanced (relative to the baseline 200 msec condition). Figure 2 shows just the opposite result; the effect of task load factors increased across acquisition and this conclusion was supported by a statistically significant interaction ($p < .01$) of acquisition days X frame duration. Thus, none of the age groups tested in this study seem to have developed automatic detection responses as evidenced by decreasing responsivity to the task load factor of display duration.

Figure 3 presents the signal detection data for the three age groups during the eight reversal days (days 11 through 18 of the present experiment). This figure is organized the same as Figure 2, and each point is based on the same number of trials. Overall, the reversal of the consistent mapping conditions had very negative consequences for performance. All three age groups showed large declines in target detection between the last day of acquisition (day 10) and the first day of reversal (day 11). These decrements were largest in the 600 and 200 msec conditions although similar decrements appear in the 100 msec condition. These large negative transfer effects provide some evidence that automatic detection processes were indeed developing across the course of the experiment. However, each of the three age groups showed comparable decrements across the reversal break--again arguing against an age-related difference in the development of automatic detection responses. A prominent feature of the Reversal Days detection performance is that it improved much less across practice than was true for the Acquisition Days condition. This outcome also argues for the development of automatic detection responses during the first phase of the experiment and suggests that those responses, once developed are difficult to eliminate in all age groups. The only exception to this conclusion is for the middle-aged subjects in the 600 msec condition. These subjects appeared able to overcome the negative consequences of previously acquired automatic detection responses under the long display duration conditions. Further research will be required to determine the extent to which this performance is typical of middle-aged adults or, perhaps unique to this sample who seemed to have exceptionally high motivation.

The results of research on visual attention reported here suggest that aging is associated with a general decrease in visual attention resources. The general decline in attentional resources seem to appear after the age of 45 and to be of equal magnitude in single frame and multiple frame visual search tasks. Furthermore, the results of the multiple frame vigilance task suggest that similar age-differences in processing resources exist in both controlled search and automatic detection modes of processing. Finally, the results of the single frame search task suggest that the visual attention deficits of the older subjects do not result from an increase in distraction created by adjacent items in the visual field.

What is the cause of these age-related differences? It is well known that cross-sectional research designs fail to provide a clear separation of cohort and ontogentic effects. The possibility exists that age differences found in the present study result from only cohort factors. What cohort factors could explain the present results? It is possible that the age groups in these investigations suffered from different childhood diseases, or may have experienced different levels of nutrition during critical stages of biological development; factors that we could expect to lead to differences in neurological functioning. It seems relatively unlikely that sociocultural factors, such as education or experience could explain the highly stable age-differences seen across the extended periods of practice administered in the present research.

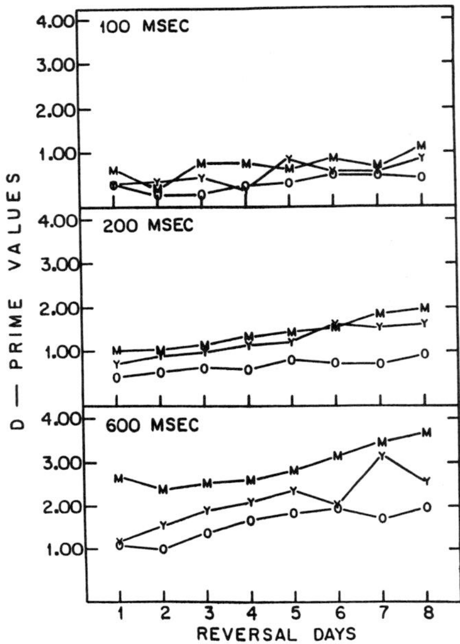

Fig. 3. Target detection sensitivity for young (Y), middle-aged (M), and old (O) subjects in a 20-frame visual search task after the original consistent mapping conditions were reversed. Eight days of practice on the reversed task are shown at three display durations.

The reported goal of most aging research is to identify ontogenetic factors. What developmental factors (assuming for the moment that we have no cohort confounds) could explain these age-differences. It seems most plausible to turn, again, to causal factors associated with neurological functioning to explain these differences. Perhaps an age-related decline in cardiovascular functioning has created a blood flow insufficiency in critical, attention mediating structures of the older brain.[14] Similarly, we might speculate that age-related degeneration in attention mediating neurons in the aging brain have led to general performance decrements.[11] Again, it seems implausible to think that these general attentional deficits result from socio-cultural variables.

The present research, unfortunately, provides no information about the underlying neurological structures that are associated with the general attention deficit seen in the older subjects. Future research could contribute significantly to our understanding of cognitive aging and its underlying causes by collecting information on neurological functioning and attentional resources from the same population of subjects. Some of the most interesting questions could be investigated with adults ranging in age from 45 to 75 years. The results of cognitive aging research show an increase, within this age range, in between subjects performance variability. Future research should examine the degree to which neurological functioning predicts attentional resources within this age range.

REFERENCES

1. Abel, M., 1972, The visual trace in relation to aging, Ph.D. dissertation at Washington University.
2. Averbach, E., and Coriell, A.S., 1961, Short-term memory in vision, *Bell Systems Technical Journal*, 40:309.
3. Eriksen, B. A., and Eriksen, C. W., 1974, Effects of noise letters upon the identification of a target letter in a nonsearch task, *Perception & Psychophysics*, 16:143.
4. Eriksen, C. W., and Colegate, R. W., 1971, Selective attention and serial processing in briefly presented visual displays, *Perception & Psychophysics*, 10:321.
5. Eriksen, C. W., and Collins, J. F., 1969, Temporal course of selective attention, *Journal of Experimental Psychology*, 80:254.
6. Eriksen C. W., and Collins, J. F., 1967, Some temporal characteristics of visual pattern perception, *Journal of Experimental Psychology*, 74: 476.
7. Eriksen, C. W., Hamlin, R. M., and Daye, C., 1973, The effect of flanking letters and digits on speed of identifying a letter, *Bulletin of the Psychonomic Society*, 2:400.
8. Eriksen, C. W., and Hoffman, J. E., 1972, Temporal and spatial characteristics of selective encoding from visual displays, *Perception and Psychophysics*, 10:321.
9. Eriksen, C. W., and Hoffman, J. E., 1973, The extent of processing of noise elements during selective encoding from visual displays, *Perception and Psychophysics*, 14:155.
10. Eriksen, C. W., and Spencer, T. J., 1968, Visual search under conditions of very rapid sequential input rates, *Perception and Psychophysics*, 4:197.
11. Freed, D., Corkin, S., Growden, J., and Nissen, M., in press, Behavioral and neurochemical evidence of locus coeruleus pathology in a sub-group of patients with Alzheimer's disease, *Neuropsychologia*.
12. Haber, R. N., and Standing, L. G., 1969, Direct measures of short-term visual storage, *Quarterly Journal of Experimental Psychology*, 21:43.
13. Haber, R. N., and Standing, L. G., 1970, Direct estimates of apparent duration of a flash., *Canadian Journal of Psychology*, 24:216.
14. Horn, J. L., and Risberg, J., 1986, Blood flow in the brain and the adulthood aging of cognitive functions, *in*: "Theoretical Empiricism: A General Rationale for Scientific Model-Building, H. Wold, ed., Paragon House, Washington, DC.
15. Kline, D. W., and Baffa, G., 1976, Differences in the sequen tial integration of form as a function of age and inter stimulus interval,*Experimental Aging Research*, 2:333.
16. Kline, D. W., and Orme-Rogers, C., 1978, Examination of stimulus persistence as the basis for superior visual identification performance amongst older adults, *Journal of Gerontology*, 33:76.
17. Kline, D. W., and Schieber, F. J., 1981, What are the age differences in visual sensory memory?, *Journal of Gerontology*, 36:86.

18. Kline, D. W., and Schieber, F. J., 1982, Visual persistence and temporal resolution, in: "Aging and Human Visual Function," R. Sekuler, D. Kline, & K. Dismukes, ed., Alan R. Liss, New York.
19. Neisser, U., 1967, "Cognitive Psychology," Appleton-Century-Crofts, New York.
20. Salthouse, T. A., 1976, Age and tachistoscopic perception, Experimental Aging Research, 2:91.
21. Schneider, W., and Shiffrin, R. M., 1977, Controlled and automatic human information processing: I. Detection, search, and attention, Psychological Review, 84:1.
22. Shiffrin, R. M., and Schneider, W., 1977, Controlled and automatic human information processing: II. Perceptual learning, automatic attending, and a general theory, Psychological Review, 84:127.
23. Sperling, G., 1960, The information available in brief visual presentations, Psychological Monographs: General & Applied, 74:1.
24. Walsh, D. A., 1982, The development of visual information processes in adulthood and old age, in: "Aging and Human Visual Function," R. Sekuler, D. Kline, and K. Dismukes, eds., Alan R. Liss, New York.
25. Walsh, D. A., and Prasse, M. J., 1980, Iconic memory and at tentional processes in the aged, in: "New Directions in Memory and Aging," L. W. Poon, J. L. Fozard, L. S. Cermak, D. Arenberg and L. W. Thompson, eds., Lawrence Erlbaum Associates, New Jersey.
26. Walsh, D. A., and Thompson, L. W., 1978, Age differences in the persistence of visual storage, Journal of Gerontology, 33:383.

DIAGNOSIS (AND DIAGNOSTIC NEEDS) OF OLDER LOW VISION PATIENTS

Marcus D. Benedetto

Ophthalmic Research Center, Orlando, Florida, U.S.A.

Low vision is that nebulous area between normal sight and total blindness. Modern medicine deserves credit for contributing to the significant reduction of total blindness. Unfortunately, in many cases this leaves the patient with some residual visual impairment. The patient is not restored to total sight. There are several definitions of low vision.

The two most common are clinical and societal. With regard to functioning within a society, low vision may be viewed as any visual impairment which interferes with the acquisition of the most appropriate education, employment, or meaningful recreation. This means that for each individual there may be a separate and unique set of factors which may inhibit that individual from appreciating his/her life and realizing her/his potential. The visual characteristics required for playing jai-alai, watchmaking and setting bowling pins are each different and are each sensitive to a different type of visual loss.

Clinically, low vision is defined as either a visual acuity in the better eye of no more than 20/70 or a visual field loss of 30 degrees or less. (The legally blind are included in this group.) Unfortunately these numbers have value only in the clinical setting and are limited in being predictive of actual visual performance in the 'real world.' The patient's ability to perform in the 'real world' appears to be dependent upon other intrinsic skills, which have, heretofore, received little attention in the clinical setting.

Individuals who are afflicted with low vision often complain that they are tagged with labels that follow them in life and severely limit vocational, educational, or recreational opportunities. The label they receive, they feel, tends to advertise a restrictive ability to operate in a visual setting. This judgement is not based upon performance, but is made by "normally sighted" individuals who project how they would operate under similar situations. Such people as Genensky and Colenbrander have been active in advocating more descriptive terminology in trying to characterize individuals who have visual problems specifically in terms of what the individuals can do as opposed to describing their total situation as being a set of negatives.

Aging has been an accepted fact of life historically. It has generally been acknowledged that as chronological age increases the physical peak is past. Hopefully the loss of speed and agility is sufficiently compensated with sagacity. In days of old, when people did not live as long as they do today, the quality of other physical attributes waned before

The Effects of Aging and Environment on Vision
Edited by D.A. Armstrong *et al.*, Plenum Press, New York, 1991

vision. Hence, until recently, vision was assumed to remain relatively constant throughout one's life unless one suffered from injury or disease. However, thanks to modern medicine, people are living longer and now visual changes and difficulties with advancing age are receiving notice.

Currently, almost two-thirds of the population over the age of 65 have low vision. Fortunately, most of these have sufficient residual vision to benefit from the use of some type of low vision aid. Unfortunately, little was known about the effects of aging upon visual performance until recently. It has been a generally accepted fact that as one gets older the need for reading glasses will arise. More light to perform the same task will be needed, and so on. But it has not been delineated what is acceptable and unacceptable visual performance criteria for a given visual task for a given age bracket.

The current diagnostic techniques involve six major areas: taking a history, observation, refraction, magnification, visual fields, and auxiliary tests. The purpose of the 'history' is to obtain not only a medical history of the patient and to get a feel for previous ophthalmic problems or possibly contributory pathology, but also to gain insight into the individual patient and his/her particular vision needs. Until the onset of the difficulty was the patient an inveterate reader, or did he/she only read the Sunday headlines once a month? What are the things that the patient used to do, and how often? What would give the patient pleasure now? Playing bridge or shuffleboard or driving a car or reading the labels on cans in the supermarket have different visual requirements. These needs have different clinical implications.

Through observation the clinician may observe a variety of potential visual difficulties. Beginning with, for example, noting whether the patient came alone or accompanied. If accompanied, what is the patient's relationship to that individual? If accompanied, how much, and under what conditions does the patient rely upon his/her companion, or other people? Merely watching how the patient negotiates getting from the waiting room to the trial lane and how the patient addresses the examining chair may even prove to be highly significant. If the patient is accompanied, then more information can be ferreted out based upon the observations of that individual and/or others familiar with the patient in other than the clinical setting.

After a thorough investigative history and observation, the next step is generally a careful refraction. It is surprising how far refining the spectacle correction can go toward resolving the patient's dilemma.

A determination of the optimal degree of magnification is made based upon refractive results and patient visual performance needs. Considering, for example, whether one is faced with a near or distance task and the field of view requirements. Once the most functional image size for the task is determined, then the type of aid most suited for the specific task and most flexible for other tasks is chosen. The most common are the hand-held, stand, or head-borne aids. The advantage of the hand-held is that it is generally the most portable and least expensive. The disadvantages are that it ties up one hand and requires the patient to make an effort to maintain the appropriate viewing distance from the material. The stand magnifiers also tie up one hand but require little effort to maintain the proper viewing distance. These tend to be more bulky.

The third class, head-borne, have several variations. They can vary from a strong pair of reading glasses to a headband mounted loupe, telescope, or microscope. These may be flipped in or out or view as needed. Loupes may be clipped to a pair of spectacles. More sophisticated bioptic spectacles with a built-in telescope or microscope may be utilized. The major advantages of this type are that it frees the hands and that the range of variety and complexity avialable to meet the patient's needs is greatest. The disadvantages are that they generally cost more and are sometimes cosmetically unattractive.

The quality and quantity of the field of view available to the patient plays an important role in readapting the patient to his/her environment. It is at the periphery of the visual field where the majority of the critical size, shape, distance, and motion judgements are made. The visual field does not usually receive much attention clinically since the traditional emphasis is generally upon the central few degrees of vision. Despite the fact that most low vision patients have disturbances of the macular and foveal region, most visual therapy is still directed to the continued delivery of magnification to this area. This is because the major drawback of the periphery is that it has poor color discrimination and does not excel at fine detail resolution. So it has been traditional to apply magnification to cover more of the central visual field with an enlarged stimulus in hopes that the area formerly noted for its high resolution ability may be able to perform adequately with a larger image despite the incurred impairment. With magnification, there is a trade-off. When magnification is employed, the field of view suffers reduction as the amount of magnification is increased. This may or may not be beneficial, depending upon the circumstances.

Auxiliary tests, such as color, contrast sensitivity, stereopsis, etc. are sometimes employed. These and other auxiliary tests tend to be under-utilized. When used, they tend not to be exploited to their fullest potential. For example, a color test may be given to determine the extent of the patient's color vision. The clinician may even sample the patient's vision through a handful of different filters. But, to date, it is a rare event when the color vision test results are strongly related to patient performance with various filters and even rarer when both of these results bear any more than a vague resemblance to the final tint in the patient's spectacles. Most of the time the spectacle tint chosen is for cosmetic value not for enhancing visual performance.

There are some very strong difficulties in diagnosis: As mentioned above, it is sometimes very difficult to determine if the behavior observed in the clinic is an appropriate and representative sample of patient performance (and difficulties) under 'real world' visual conditions. The clinical trial lane is certainly a poor simulation of 'real world' visual conditions. Hence, it should be reemphasized that the limitation of observation is the artificiality of the clinical situation.

Additionally, the clinician is faced with the discrepancy between the patient's current and remembered perceptions of the world. Even the most idyllic image reviewed under ideal conditions may not match the patient's internalized remembrance. It can be difficult to delineate realistic desirable goals.

Moreover, the confounding variable of age wreaks the most havoc. It is very difficult to determine what are and to what extent are normal changes with age for this individual or bracket of individuals. How much of the visual difficulty is due to normal aging and how much is abnormal, deserving of more critical investigation? Despite the fact that it is now well known that virtually every aspect of the eye and vision changes with age, there are currently no standards for 'normal' vision ranges for a given age or age bracket.

The exclusive use of refraction and magnification should not remain the panacea they have become. Galileo's invention of the telescope,approximately 300 years ago, was first employed to help compensate for visual problems. Little has changed since. The same basic principles are being used - magnification in the central few degrees, merely the style of the carrier frame has been changed. It is, essentially - "old wine in new bottles." Magnification is still the primary mode of choice, for most clinicians, in dealing with the visually impaired. Unfortunately the use of magnification involves a tradeoff of valuable, under-utilized visual field for increased image size in the central few degrees of vision. For many patients magnification does not clear up the image and make it sharp and distinct, only bigger. More investigation should be made into the use of the peripheral vision which is relatively unaffected by refractive error in contrast to the central vision which is highly sensitive to errors of focus.

As alluded to earlier, the auxiliary tests have remained relatively unaddressed. Color vision tests could be used to determine the optimal spectacle tint to enhance visual performance. Contrast sensitivity could be used to get a better concept of 'real world' visual performance. The standard acuity test is done under only 100% contrast - usually black letters on a white background. However, the 'real world' is not black and white but less than 100% contrast. If only 100% is tested, it is difficult to know how the patient will perform in anything other than the ideal clinical situation.

Despite the fact that almost all non-clinical tasks are dynamic to some degree, motion is virtually never addressed clinically. Static visual performance has been shown to have little relationship to dynamic visual performance. (1) This is especially true under degraded image conditions and even truer in the peripheral field of vision.

The majority of the patient population has, for the most part, based the majority of their experiences upon the kind of visual resolution that the five degree central field makes possible. Characteristics of the peripheral visual field which might be utilized for visual judgements have been either ignored or under-utilized. If one were to regard this from a developmental sense, it would seem that the peripheral visual mechanisms are older and have the greatest survival value. In terms of comparative anatomy, physiological adaptation seems to require a peripheral visual system almost all low vision patients possess. Only a small number of elderly low vision patients have peripheral visual field involvement. Almost all have central field involvement. So, in terms of new directions, the exploration of how to utilize literally the remaining 160 degrees of visual field as opposed to systems that attempt to compensate for simply the loss of a five degree central field is strongly recommended.

Therefore, the task of clinicians and visual scientists is to find means of utilizing the characteristics that do exist in the peripheral field. The patient should be exposed to training, biofeedback and other techniques of increasing individual awareness of these mechanisms in terms of developing a systems approach that the individual can utilize to make appropriate adequate visual judgements. To this end, study and research is recommended in the utilization of techniques to enhance edge contrast, color, motion detection, and other visual performance characteristics to reduce the impairment in visual resolution. The exploration of potential in devising spatial and chromatic shifts which may increase the clues available to the impaired visual system for correct and adequate visual judgements should be encouraged.

This view of low vision diagnosis and therapy is not the traditional present medical model which essentially delivers therapy on a 'crisis intervention' basis; but rather that here is an individual who requires visual rehabilitation in order to meet the visual demands of the real world. This cannot be a "limp in - leap out" one visit type of approach, but truly requires rehabilitation and long term adaptation. In short, vision rehabilitation should be a matter of visual re-education. We are encouraged in our belief that this can be done by recent results which have been observed in the treatment of closed head injuries and strokes where patients were seen very early in the course of their treatment and appropriate visual therapy applied. Visual fields which appeared to have been irretrievably lost were demonstrated to have reversal and improvement. (2)

There does seem to be evidence for plasticity (3) in the visual system and the potential for utilizing mechanisms previously ignored for the development of a support system for the given individual in meeting the needs of his/her environment. There is a simultaneous need to develop more adequate, more sophisticated tests of visual functions and the development of devices which utilize these findings in re-educating the patient. An opportunity truly exists to minimize the impact of age and disease upon the visual system.

SUMMARY

Patient needs and dissatisfaction with current diagnosis and treatment demand that low vision depart from its 300-year-old tradition and embark upon a new direction: 1) The exploration of the enhancement of residual visual function without focussing solely upon magnification. 2) Exploring auxiliary tests and culling from them the fullest extent of information. 3) Utilizing this new information to develop appropriate therapies to enable the patient to learn to enhance his/her visual performance, despite his/her subnormal vision. 4) Developing improved support systems at home, work, and play for the patient. 5) Recent vision rehabilitation work supports the notion that visual losses are not immutable there is plasticity available to be tapped to reverse visual damage. This controverts the standard medical model- "if it's broke, it's broke!" 6) Diagnostic needs for more rigorous investigation of personal needs, wants and desires yoked with closer to real life environmental clinical correlation. 7) Age-related standards and norms for visual performance. 8) Better tests need to be designed to give the clinician more useful, functional information. 9) Better therapies to more closely meet the patient's needs and more tightly tied to diagnoses and tests.

One of the greatest difficulties in dealing with the low vision patient is that low vision is not one set type of visual defect, but encompasses a myriad of visual impairments with a myriad of variations on a series of continuums anywhere between normal sight and total blindness. The truly confounding problem in dealing with low vision is that even for any two patients with clinically identical visual impairments at the same level on the same continuum for that particular condition, their perceptions of the world may be as different as night and day. Unfortunately, the problem of dealing with low vision is compounded by changes with age. Until recently clinicians have concentrated upon the central few degrees of vision employing magnification as their major tool. It is, therefore, essential that the above nine points be pursued in order to augment the armamentarium of the low vision practitioner; thus improving his/her effectiveness in enhancing the visual performance and, consequently, the plight of the low vision patient.

BIBLIOGRAPHY

1. Benedetto, M.D., and Gaynes, E.M., 1986, The effect of image degradation upon visual performance thresholds. Presentation at the American Academy of Optometry. Toronto, Canada, December.

2. Benedetto, M.D., 1986, Low vision revisited: Vision rehabilitation: present and future trends. McFarland Seminar, presented at the International Association for Education and Rehabilitation of the Blind and Visually Impaired Conference. Chicago, Illinois, July.

3. Bridgeman, B. and Stagg, D., 1982, Plasticity in human blindsight. Vision Res. 22, 1199-1203.

MULTIDISCIPLINARY ASSESSMENTS OF VISUAL AND PSYCHOMOTOR FUNCTIONS ESSENTIAL FOR DRIVING IN THE ELDERLY

J.M. Ordy[1], G.J. Thomas[2], T.M. Wengenack[2], W.P. Dunlap[3], and R. Kennedy[4]

1. Fisons Pharmaceuticals, Rochester, NY
2. Univ. of Rochester, Rochester, NY
3. Tulane U., New Orleans, LA
4. Essex Corp., Orlando, FL

VISION, AGING, DRIVING: PERSONAL, SOCIAL, ECONOMIC, LEGAL, AND SCIENTIFIC PERSPECTIVES

It appears intuitively clear that driving a motor vehicle involves complex visual and psychomotor skills. Driving a vehicle may be the preferred mode of transportation/mobility on urban and rural roadways. This transportation/mobility may be a very important factor in personal, social, and economic independence for individuals of all ages, including the elderly. Even a brief and selective review of the extensive basic and applied research literature on vision, aging, and driving clearly indicates the following: 1) the increasing proportion of the elderly in the total population may need or prefer driving independence or transportation/ mobility as long as they appear to drive safely, 2) there is evidence that, as part of normal aging, visual and motor skills necessary for day, and particularly for night driving may begin to decline by age 55 or 65, and then more noticeably by age 75, 3) various visual and motor driving skills, such as acuity, contrast and glare sensitivity, peripheral fields, divided attention, reaction time, and many others, may show different stages of onset, or rate of decline, within individual elderly, as well as among different elderly subjects, 4) visual impairments may be related directly to such ocular anomalies or pathologies as errors of refraction, cataracts, glaucoma, binocular field defects, or age-related macular degeneration (AMD), 5) elderly drivers may recognize many visual-motor limitations and unsafe driving practices, and take some remedial actions, 6) recognized, or unrecognized visual and motor skill problems of the elderly may result in a preponderance of certain types of driving safety problems, traffic violations, and accidents, and 7) basic and applied research may help identify visual and motor skill problems in the elderly that may be amenable to correction and improved driving performance, better design of vision standards, improved safety, and lower accident rates.

Population projections indicate that the proportion of elderly of the total population, as well as the proportion of elderly drivers will continue to increase dramatically in the USA and other industrialized countries. Although the importance of visual and motor skills essential for driving are intuitively recognized, it is only recently that there has been an increasing scientific awareness that age-related changes in visual and motor skills, increased ocular pathologies, and probably some, as yet unrecognized, visual and psychomotor problems may also be associated with

The Effects of Aging and Environment on Vision
Edited by D.A. Armstrong *et al.*, Plenum Press, New York, 1991

problems of driver safety and accident rates among the elderly drivers. Whereas population trends indicate that the proportion of elderly drivers will continue to increase and become a major problem for basic, clinical, and applied research, as well as for social policy planners, and regulatory agencies, there are also some misleading stereotypes of the elderly such as being largely residents of high density urban living areas, and largely dependent upon public transportation. According to recent surveys, however, a majority of the elderly over 55, may live in suburban or rural communities, have a current driver's license, own cars, and use them as the preferred mode of transportation/mobility (Rosenbloom, 1988). Although the specific aim of this brief and very selective review has been to focus on multidisciplinary test assessments of visual and psychomotor functions essential for driving in a "mixed-age" population of drivers, perhaps an equally important aim has been to highlight possible basic, clinical, and applied research opportunities for optimizing visual and driving performance functions of the elderly by focusing on more appropriate tests, increased awareness, retesting, training, or possible modifications of driving standards, traffic signs, as well as vehicle and roadway designs. Specific aims of this review of vision, aging, and driving, have been to highlight, and examine in some detail, potential research problems in the following areas: 1) scope of the problem of elderly drivers, 2) general vision standards, 3) use of the vision standard of photopic 20/40 static visual acuity (SVA) to predict dynamic acuity and other visual functions actually used under day or night driving conditions, 4) acuity, contrast, adaptation, and glare as problems for night driving in the elderly, 5) age differences in relationships among static and dynamic visual functions essential for driving, 6) age differences in the eye and the visual pathways and centers of the brain involved in vision and driving, 7) incidence of ocular anomalies, pathologies, and ophthalmological problems among the elderly with possible limiting effects on driving, 8) driving by the elderly in context of other human factors, 9) safety problems, traffic violations, and accident rates among the elderly, 10) elderly driving performance under urban, highway conditions, traffic signs, and vehicle designs, and 11) research opportunities for optimizing visual and driving performance of the increasing proportion of elderly drivers in the total population of drivers.

SCOPE OF PROBLEM: DYNAMIC CHANGES IN PROPORTION OF ELDERLY AND ELDERLY DRIVERS IN THE TOTAL POPULATION

One of the major demographic changes in the USA during this century has been the dramatic increase, both in number and proportion, of the elderly in the total population. According to estimates of recent and prospective trends of the "age-mix" of the total population, there were 3.1 million people in the USA over age 65 in 1900, representing 4.1% of the total population. By 1980, there were more than 23 million over 65, representing 11% of the total population. It is estimated that by the year 2020, at least 16% of the total population of the USA will be over 65 (Siegel, 1980). In terms of the proportion of elderly drivers in the total population of drivers, estimates indicate that the proportion of elderly drivers will continue to increase markedly with the increased proportion of the elderly, in the total population. According to some recent estimates, there are approximately 152 million licensed drivers in the total population, with approximately 33 million drivers of age 55 and over, representing 22% of all drivers. By the year 2000, they will represent 28%, and by 2050, 39% of the total driving population (Malfetti, 1985). The rate of increase of female drivers is greater than that of male drivers. Basically, while the number of teenage drivers will decrease as the peak of the "baby boomers" passes to middle age, the number of elderly drivers will be increasing dramatically. In a broader context, it has been estimated that there may be 2 to 6 million people in the USA with "poor" vision, and that over 500,000 may be drivers with acuities ranging from 20/100 to 20/175 (Fonda, 1986; see Table 1 of this review for minimum acuity standards for a driver's license among states).

Although there are probably plausible but debatable reasons for defining the "elderly" driver in terms of functional visual and motor skill characteristics essential for driving, more commonly, some chronological age, i.e., 55 or 65 years, is used as a defining characteristic of the "elderly" driver. It seems interesting to note, however, that as the proportion, or "age-mix" of the elderly in the total population increases, functional and chronological criteria for defining the "elderly" driver are also likely to change (Birren and Williams, 1982). Because there is some recognition that functional, or biological, and chronological age may become "dissociated" with increasing age, drivers 55 and over may comprise a significantly greater range of visual and motor skills, compared to young, or middle aged groups. Consequently, there is concern, particularly among many elderly drivers that, restrictions or withdrawal of a driver's license should not be jeopardized solely because of chronological age (Malfetti, 1985).

GENERAL VISION STANDARDS OR REQUIREMENTS FOR DRIVING IN A CHANGING "MIXED-AGE" POPULATION

The importance of vision for driving is well recognized by individuals, policy planners, regulatory agencies, and human factors' engineers. All states in the USA, as well as the District of Columbia and Puerto Rico, require driver's license applicants to pass some type of visual screening test, usually static visual acuity (SVA) under high illumination, at least to obtain their first driver's license. Subsequently, however, there is little standardization of visual tests, or test procedures, minimum standards, or requirements, and frequency of visual retesting with increasing age, among the different states. Forty-one of the states require visual tests at the time of driver's license renewal, but there is great variation in the time intervals between renewals, test-relevance, special tests for the elderly, and enforcement of the visual standards or requirements. Although driving involves dynamic visual acuity (DVA) and other dynamic visual functions, only SVA tests are required. There are wide variations in SVA requirements among states, ranging from 20/30 to 20/60, in the better corrected eye. Only 19 states routinely require some type of visual field tests, and 14 other states require visual field tests for commercial, or special driver's license holders. Many studies have focused on correlations among visual field defects, driving performance, traffic violations, and accident rates (Keltner and Johnson, 1987). Drivers with visual field defects in both eyes have been reported to have traffic accident and conviction rates twice as high as age- and sex-matched controls with normal visual fields. Perhaps more significant, 60% of individuals with visual field defects were previously unaware of problems with their peripheral fields of vision. Interestingly, drivers with visual field loss in only one eye did not differ significantly in accident and conviction rates from age- and sex-matched controls with normal visual fields (Keltner and Johnson, 1987). It seems likely that as a starting point in basic and applied research on vision, aging, and driving, these minimum vision standards, tests and procedures, could provide a framework for adjusting or expanding vision screening upon renewal of a driver's license with increasing age. Thus far, very few states have conducted comprehensive surveys or scientific studies of visual functions and driving performance in the elderly. Only 7 states have specifically focused on vision and driving in the elderly, and 20 states have begun to accumulate statistical data on moving violations and accident rates for different age groups (Keltner and Johnson, 1987). Table 1 provides a summary of survey results of visual

Table 1
Visual Test Procedures and Requirements for Driving Motor Vehicles in 50 States (plus Puerto Rico and the District of Columbia)*

1) Any Type of Applicant Screening by Visual Tests; Number of States:
   Yes: 52   No: 0

2) Screening by Visual Tests for License Renewal; Number of States:
   Yes: 41   No: 11

3) Screening by Special Tests According to Age**; Number of States:
   Yes: 3   No: 49

4) Static Acuity Requirements (better corrected eye); Ranges Among States:
   20/30: 8   20/40: 39   20/50: 2   20/60: 3

5) Tests of Binocular Fields; Number of States:

Routine for all license applicants:	18
Never tested:	28
Professional drivers only (e.g. bus driver):	4
For license renewal only:	1
No Report:	1

* From Keltner and Johnson (1987; Table 1, p. 1182)
** Vision screening upon license renewal in designated age groups.

screening standards, or requirements for driving a motor vehicle in the 50 states, Puerto Rico, and the District of Columbia.

VISION STANDARD OF 20/40 STATIC VISUAL ACUITY (SVA) UNDER "NORMAL" DAYLIGHT (PHOTOPIC) ILLUMINATION AS PREDICTOR OF TWILIGHT (MESOPIC) AND NIGHTTIME (SCOTOPIC) DRIVING CONDITIONS

Static visual acuity (SVA) has been used extensively as a measure of the visual system's capacity to resolve spatial detail, as a clinical procedure for evaluation of errors of refraction, as a functional starting point for studying basic optical and neural mechanisms of spatial vision, and as a general screening procedure in a variety of visually-dependent performance tasks (Keltner and Johnson, 1987; Ordy, et al., 1991). Some of the more common tests for evaluation of SVA include the Snellen chart, Landolt rings, and a variety of checkerboard targets (Pitts, 1982). The results of such SVA tests are generally expressed as a fraction or decimal, and refer to the "minimum angle of resolution" (MAR) of the visual system to resolve, or discriminate objects in the visual environment. As indicated previously, there is no single minimum standard of SVA among the 52 districts, but 39 of the 52 districts (see Table 1) require an acuity score of 20/40, using the better corrected eye, obtained under "normal" (photopic) illumination, for receiving a driver's license (Federal Highway Administration, 1986). In addition to the presumed validity of SVA as a predictor of dynamic visual acuity (DVA) that is obviously involved in driving a motor vehicle, there is a very practical appeal for use of static acuity screening tests for obtaining a driver's license. The acuity tests are brief, relatively simple, and quite well understood as a basic screening procedure for the ability to discriminate spatial detail (Federal Highway Administration, 1986). Because minimum standards of visual acuity vary considerably among states, Table 2 provides a brief summary of "normal" adult 20/20 acuity, and the minimum acuity standards for different states (see Table 1), expressed as acuity equivalents in "minimum angle of resolution" (MAR), Snellen decimals, fractions, and Log MAR.

Table 2
Minimum Visual Acuity Standards Among States Expressed as Acuity Equivalents*

Snellen		Minimum Angle Resolution (MAR)**	Log MAR***
Fraction	Decimals		
20/20	1.00	1.00	0.00
20/30	0.67	1.50	0.18
20/40	0.50	2.00	0.30
20/50	0.40	2.50	0.40
20/60	0.33	3.00	0.48
20/70	0.29	3.50	0.54
20/80	0.25	4.00	0.60

* See Table 1 for minimum acuity standards among states.
** Minimum angle of resolution expressed in minutes of arc.
*** Log MAR transformations for statistical evaluations.

The apparent validity of SVA as a minimum standard and as a predictor of DVA essential for driving a motor vehicle is based on assumptions that: 1) there is a significant correlation between measures of static and dynamic acuity, 2) that both measures of acuity remain relatively constant as functions of illumination and age, and 3) that the proportion of elderly drivers of the total driver population also remains constant and relatively negligible. Although measures of SVA appear brief, simple to administer, and easy to interpret in terms of minimum acuity standards for driving a motor vehicle, these SVA measures are known to be systematically dependent upon luminance, contrast, and duration, as well as to a variety of optical and neural properties of the eye and the geniculostriate system of the brain (Ordy, et al., 1991). Although earlier studies indicated that "normal" (photopic) acuity of 20/20 may remain relatively constant from 20 to 60 years of age (Pitts, 1982), an increasing number of studies have indicated that even normal photopic SVA may decline significantly with age as a function of lower illumination and contrast (Richards, 1977; Pitts, 1982; Adams, et al., 1988; Sturr, et al., 1990). In one recent study, 60 young

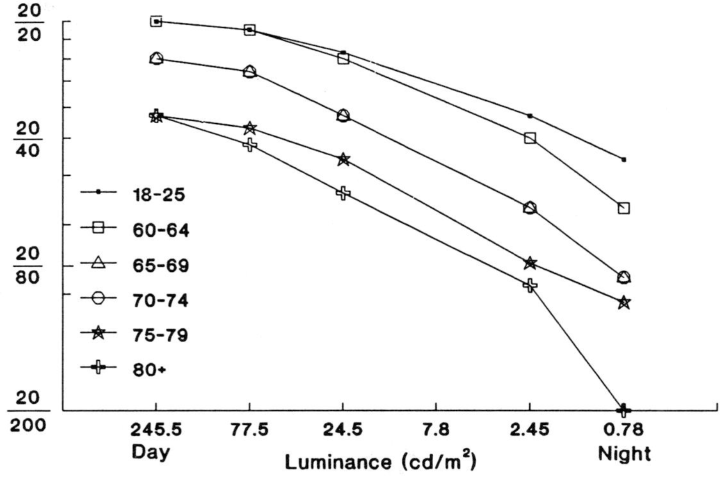

Figure 1
Visual Acuity as a Function of Age and Luminance Snellen Fraction

(18-25 years) and 91 elderly (60-87 years) volunteers were tested for SVA under six levels of illumination, ranging from 245.5 (photopic) to 0.2 (scotopic) cd/m^2 levels. The elderly were subdivided into 5, half-decade subgroups: 60-64, 65-69, 70-74, 75-79, and 80 years and above. The decimal acuity scores of each subject were first transformed logarithmically for statistical comparisons of acuity differences as functions of age and luminance (see Table 2). Statistical comparisons of acuity differences among six age groups (including 18-25 years) across 5 illumination levels indicated significant age differences among groups with decreasing levels of illumination (Sturr, et al., 1990). Although these important experimental findings can certainly be challenged because of the relatively small number of subjects in the 5 half-decade subgroups from 60 to 80 years and above (60-64: n=13; 65-69: n=29; 70-74: n=21; 75-79: n=20; 80+: n=8), Figure 1 illustrates the significant decrease in visual acuity as functions of age and luminance (based on data in Figure 1, p. 4, Sturr, et al., 1990).

In terms of specific age-group comparisons, however, it seems important to note that there were no significant differences in acuity as a function of luminance between the young (18-25) and older (60-64) age groups, even at the relatively lower (night) level (0.78 cd/m^2) of illumination. In contrast, even at the higher, photopic levels of 245.5 cd/m^2 and 77.5 cd/m^2 of illumination, elderly of half-decade subgroups 75-79 and 80-87 years had significantly lower acuity compared to the young (18-25) and the 60-64 years age groups (see Figure 1).

As indicated previously, a 20/40 acuity represents a minimum standard for screening drivers in 39 of 52 districts (see Table 1). In view of the significant decrease in acuity as functions of age and luminance, particularly above 65 years, it appeared important to evaluate the percentage or proportion of drivers passing a minimum 20/40 standard as functions of age and luminance level. This evaluation can be illustrated graphically for the same 60 young (18-25) and 91 elderly (60-87 years) subjects, as described in the previously cited study (Sturr, et al., 1990, Table 1, p. 5). Figure 2 illustrates the percent of drivers passing a minimum 20/40 acuity standard as a function of 6 luminance levels. In order

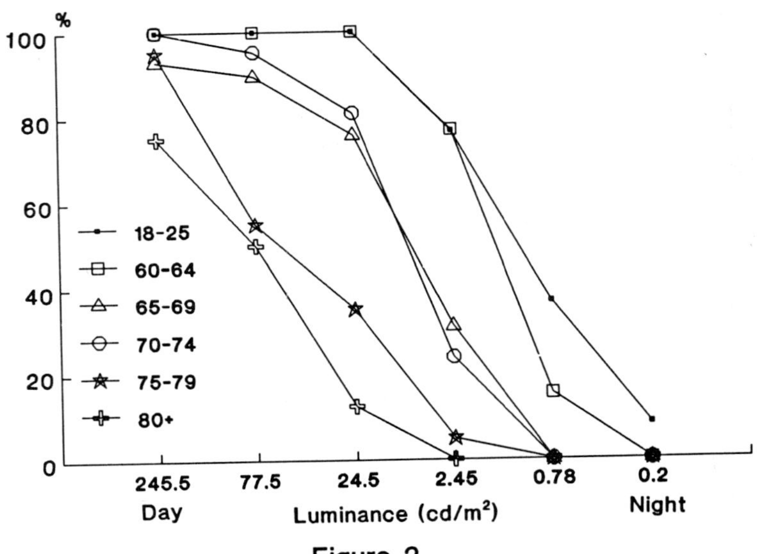

Figure 2

Percent Passing 20/40 Acuity Standard as Function of Luminance

to make age comparisons of acuity and luminance levels more meaningful in terms of day-night driving condition, the 245.5, 77.5, and 24.5 cd/m^2 levels represent daytime, or photopic, luminance, the 7.8 and 2.45 cd/m^2 levels twilight, or mesopic, and the 0.78 and 0.2 cd/m^2 levels within the night driving (scotopic) luminance range (Richards, 1977; Sturr, et al., 1990). As shown graphically in Figure 2, the percent passing a 20/40 acuity standard remained relatively constant from 18-25 to 60-64 years at the 245.5, 77.5, 2.45, and even the 0.78 cd/m^2 luminance levels, whereas the pass/fail frequencies of the 65-69, 70-74, 75-79, and 80+ age groups decreased significantly from the 24.5, 2.45, 0.78, to the 0.2 cd/m^2 luminance levels (see Figure 2). In general, the cited study (Sturr, et al., 1990), with the results on visual acuity as functions of age and luminance, and the percent passing 20/40 acuity as function of age and luminance, illustrated in Figures 1 and 2 respectively, appear in general agreement with other studies, suggesting that many elderly over the age of 60-65 years may experience greater declines in static visual acuity (SVA) compared to younger age groups under decreasing levels of illumination and contrast, particularly under scotopic nighttime driving conditions.

Some new scientific perspectives are beginning to emerge concerning visual acuity differences or changes with age, luminance, and spatial contrast sensitivity, with possible implications for the use of 20/40 SVA performed at "normal", medium photopic levels of illumination, as a predictor of driving by the elderly under DVA conditions, lower illumination, and contrast (Sekuler and Owsley, 1982; Adams, et al., 1988). In terms of age differences in photopic acuity, it is still widely accepted that "corrected" eye Snellen chart acuity of 20/20 remains relatively constant from 20 to 60 years of age (Pitts, 1982). However, several factors have been proposed that may challenge, at least in part, this relatively plausible but possibly too broad generalization concerning even photopic acuity and age. The conventional high contrast (85-95%) Snellen chart is unlikely to result in acuity measurements that are also sensitive to reductions in contrast sensitivity in the elderly. Young adult photopic acuity may be considerably better than the "normal" value of 20/20. Many acuity charts may not have letters or targets corresponding to higher acuity levels. If assumptions are made that photopic 20/20 acuity is the young adult "norm", attempts may not be made to record acuity that is higher than this norm. As a consequence of the extensive use of cross-sectional designs, the "artificial 20/20 ceiling" based on the optimum adult photopic acuity norm, and the increasing variability of this acuity in progressively older cross-sectional age groups, basic and clinical studies of photopic acuity and aging have thus far only grossly estimated the onset, magnitude, and rate of this acuity decrease in otherwise ophthalmologically normal subjects, from maturity to senescence (Ordy, et al., 1991). It could be argued that, given the optimum photopic acuity of 20/20 or better in young adults, and the much lower minimum vision standard of 20/40 among most states, that the relatively low minimum acuity standard assures issuance of a driver's license to applicants to at least age 60 and over, when photopic acuity begins to decline at a more rapid rate (Pitts, 1982; Ordy, et al., 1991). Future studies of photopic acuity, aging, and driving may focus on correlations between age-related decreases in photopic acuity (20/20?), or the 20/40 minimum standard, and various measures of driving skills of the elderly, or with specific safety problems, traffic violations, or accident rates. However, with the exception of readily detectable and correctable errors of refraction by glasses, it will be very difficult to isolate age-dependent decreases in normal photopic acuity as the "major, or contributing cause", in safety problems, traffic violations, or accident rates in the elderly (Sturgis and Osgood, 1982; Keltner and Johnson, 1987).

Although the importance of photopic acuity in aging and driving continues to be emphasized, in the visual world, however, not all objects are encountered at the same size, form, color, or at high contrast. A psychophysical spatial contrast sensitivity function (CSF) has been proposed as a measure of a subject's capacity to resolve objects or targets, ranging from fine to course spatial forms, or structures, and levels of contrast, by the use of sine wave grids of varying visual frequency (Sekuler and Owsley, 1982). In the CSF test, sinusoidal gratings are used which have two parameters: 1) spatial frequency, or cycles per degree (c/deg), and 2) contrast sensitivity, defined as the ratio, in which the numerator represents the difference between the grating's highest and lowest luminances, and the denominator of the ratio which represents the sum of the grating's highest and lowest luminances. Basically, the CSF measure is based on a subject's capacity to resolve objects with different contrasts from a continuum of spatial frequencies (Sekuler and Owsley, 1982). The adult photopic acuity of 20/20 can be expressed as 1 minute of arc, or by c/deg of visual angle (see Table 2 for acuity equivalents). Classical acuity measures represent an important cutoff on a spatial frequency and contrast continuum. Relations between classical contrast sensitivity ($\Delta I/I$) and spatial CSF need to be clarified. In one study, significant age differences in low spatial contrast sensitivity were reported between normal young and a "select" group of healthy elderly subjects even though both age groups had comparable acuity with high frequency and contrast targets (Sekuler, et al., 1980). In subsequent studies, however, it was reported that with a larger sample of elderly subjects, spatial contrast sensitivity decreases were found predominantly for intermediate and high frequencies (Sekuler, and Owsley, 1982, Owsley, et al., 1983). Other studies have indicated that decrease in spatial contrast sensitivity in the elderly not only increased with increasing spatial frequency, but also with decreases in luminance levels (Sloane, et al., 1988), and with moving targets (Sekuler and Owsley, 1982). Studies of relationships among various measures of static and dynamic visual functions, including contrast sensitivity, have indicated that reading street signs while driving, even during daytime, may represent a particular difficulty for the elderly (Kosnik, et al., 1988). As indicated previously, some of these new neuroscience perspectives on vision may begin to focus on static and dynamic acuity and spatial contrast sensitivity, with specific implications for driving by the increasing proportion of elderly drivers (Ordy, et al., 1991).

VISUAL ACUITY, CONTRAST SENSITIVITY, ADAPTATION, AND GLARE AS PROBLEMS FOR NIGHT DRIVING IN THE ELDERLY

Based on individual experience or awareness, surveys, as well as basic and applied research, it is recognized that nighttime driving, with greatly reduced illumination and visibility, represents a challenge for drivers of all age levels, and particularly for elderly drivers (Kosnik, et al., 1988). As indicated previously, visual acuity decreases significantly as a function of age and luminance (see Figure 1), and the percent of drivers passing a 20/40 acuity standard as functions of age and luminance also decreases significantly (see Figure 2). In addition to the significantly reduced acuity as function of age and luminance, age-related decreases in spatial contrast sensitivity, rates of light-dark adaptation, and problems of glare, all tend to exacerbate reduced acuity, and thus make driving by many elderly a greater challenge (Sturgis and Osgood, 1982). Although cross-sectional designs of age-related differences in acuity and other visual functions tend to overestimate age-related impairments (Storandt, 1982), significant declines with normal aging have been reported for such visual functions (important for nighttime driving) as acuity, contrast sensitivity, adaptation, stereopsis (Pitts, 1982), as well as in glare sensitivity, field of vision, and accommodation (Carter, 1982). There is considerable literature suggesting that correlations among many visual functions under photopic and scotopic conditions may be quite low for individuals and for groups (Pitts, 1982). Even though there is little experimental evidence that acuity screening for a driver's license at a level of 20/40 at normal photopic illumination can predict an acceptable acuity standard for driving under twilight (mesopic) or nighttime (scotopic) conditions, only 3 states have some restriction concerning vision screening under low illumination upon driver's license renewal for individuals in older age groups (see Table 1).

In view of the increasing proportion of elderly drivers, and the greater complexity in developing tests for night vision or visual functions under low illumination, it seems useful to help identify some of the major scientific, practical, and research problems that may be important for nighttime driving by the elderly. Some of the more challenging and unresolved problems of nighttime driving in the elderly include: 1) questions of the photopic 20/40 acuity standard as valid predictor or acceptable vision standard for nighttime driving by the elderly, 2) age-differences in spatial contrast sensitivity, adaptation, and glare, with implications for nighttime driving by the elderly, 3) individual awareness of visual problems, and self-imposed restrictions for nighttime driving by many elderly, and 4) associations between nighttime driving and problems of safety and accidents in the elderly.

A number of studies have attempted to examine the effects of reduced illumination, contrast, and glare on static visual acuity in order to help establish whether a 20/40 static visual acuity standard could also serve as a standard of visual acuity acceptable for night driving in the elderly (Sturgis and Osgood, 1982; Sturr, et al., 1990). Studies have indicated that daytime acuity is a relatively poor predictor of acuity at low levels of illumination (Sivak, et al., 1981; Sturr, et al., 1990). In terms of the validity of the most commonly used 20/40 static visual acuity standard, one study reported that, at 0.78 cd/m^2, which is within the average nighttime luminance range (0.34-1.03 cd/m^2), none of 78 elderly drivers above age 65 were able to pass the 20/40 acuity standard (Sturr, et al., 1990; see Figure 2 of this review). In terms of individual awareness of an acuity problem for nighttime driving by the elderly, only 30% of the elderly in the above cited study reported being aware of impaired acuity at night, and being "selective" about their nighttime driving habits (Sturr, et al., 1990). This individual awareness among some elderly drivers of reduced acuity at night, resulting in self-imposed restrictions of nighttime driving, has also been reported in earlier studies (Richards, 1977).

Although several studies have reported that daytime acuity may be a relatively poor predictor of acuity under low illumination or nighttime driving conditions (Sivak, et al., 1981; Sturr, et al., 1990), other studies have reported that, even though static acuity decreased significantly with age, luminance, contrast, and glare, there were high correlations between visual functions, including static acuity, tested at high and low background luminances, even in the presence of glare (Sturgis and Osgood, 1982). It was concluded that a "nighttime driving" acuity standard could be adopted for the general population of drivers, including the elderly, based on measurement of visual acuity at low levels of illumination. A critical problem for adoption of a "nighttime" acuity standard would be the selection of a specific, or reasonably low luminance level (0.34-1.03 cd/m^2) at which to set the acuity standard. For example, if the 0.78 cd/m^2 low luminance level, which is within the average nighttime driving luminance range, would be selected, progressively fewer elderly over 65 years would pass the 20/40 acuity standard at this low luminance level (see Figures 1 and 2). At present, no states require such a nighttime acuity standard, but one state restricts licensees with photopic visual acuity from 20/40 to 20/70 to daytime driving only (Federal Highway Administration, 1986).

In addition to the importance of luminance for acuity during night driving, target contrast, state of adaptation, and glare sensitivity are known to be of particular relevance for night driving by the elderly (Sturgis and Osgood, 1982; Sturr, et al., 1990). Contrast sensitivity is of obvious importance for nighttime driving since both glare and low illumination reduce legibility of roadway signs in the elderly during nighttime driving (Sivak, et al., 1981). Because of reduced legibility of highway signs at night in the elderly, they are likely to have less distance and time in which to respond to information presented on roadway signs. Spatial contrast sensitivity tests have shown significant age-related differences in highway sign discriminability even under photopic conditions (Evans and Ginsberg, 1985). Significant differences have also been shown in sign recognition between daytime and nighttime driving conditions (Shinar and Drory, 1983). Although level of adaptation can be important for driving during daylight, such as when drivers enter and exit tunnels, the more common concern is with initial dark adaptation at dusk, and with more rapid changes in the state of adaptation that occur during night driving as a result of rapid changes in roadway or roadside lighting, or headlights of other vehicles. Numerous studies have shown significant age-differences in rapidity of dark adaptation and its adverse effects on visual acuity in the elderly (Pitts, 1982).

In terms of associations between nighttime driving by the elderly and problems of safety and accident rates, there is evidence that not only acuity and other visual functions, but also other sensorimotor skills may decline and pose a variety of safety problems and accident risks. Perhaps to "compensate" for declining visual function and sensorimotor skills, many elderly drive less at night, for shorter distances, at slower speeds, make fewer lane changes, and avoid rush hours and other adverse driving conditions (Retchin, et al., 1988). Although the elderly may drive fewer miles compared to other age groups, they appear to have higher accident rates for every 100,000 miles of driving, excluding those of 25 years or younger (see Figures 3 and 4). It is widely recognized that accidents increase at night. As yet, based on various state surveys, reliable statistical evidence is lacking concerning specific relationships between night vision capacity, such as acuity, rate of adaptation, and glare, and accident type and frequency among the elderly (Keltner and Johnson, 1987).

AGE DIFFERENCES IN RELATIONSHIPS AMONG MEASURES OF STATIC AND DYNAMIC ACUITY AND OTHER VISUAL FUNCTIONS IN DRIVING

Although it is intuitively obvious that driving in the "real world" involves dynamic visual acuity (DVA), or the capacity to discriminate fine spatial details of moving objects, and thus represents the "critical criterion measure" of driving performance, thus far, only static visual acuity (SVA) tests are required for driving a motor vehicle in all states (see Table 1). Some studies have shown that increasing the angular velocity of an object relative to an observer, even above a relatively low value, results in a positively accelerated decrease in acuity. Even at the relatively slow speeds of 30-40 deg/sec, the acuity for spatial detail decreases markedly compared to stationary targets (Long and Garvey, 1988). In view of the previously described decline in static visual acuity with age and luminance (see Figures 1 and 2), it appears intuitively obvious that dynamic visual acuity (DVA) would also decline with age and luminance, to an even greater degree. Several earlier, but often-cited studies reported that SVA decreased with age, and DVA decreased to a much greater extent with increasing age and velocity of the targets (Reading, 1972; Burg, 1966). Earlier studies reported that DVA correlated with driving performance to a greater extent than SVA or any other visual function (Burg, 1968). Previous studies also reported a high correlation between SVA and DVA, implying that

SVA would determine some upper limit to DVA (Burg, 1966). Other studies, however, have also reported that subjects with identical SVA can vary greatly in many DVA tasks (Long and Garvey, 1988). Although SVA and DVA both measure spatial detail, and both also generally measure targets in horizontal orientation, relevant to driving, it may not appear surprising that inconsistent correlations have been reported between SVA and DVA for many tasks, including driving, in view of the greater complexity of interacting variables involved in measurement of DVA (Long and Garvey, 1988; Scialfa, et al., 1988). SVA and DVA are both sensitive to luminance, contrast, and duration, whereas various differences in target orientation and velocity during DVA may be the source of some of the reported inconsistent correlations between SVA and DVA in various performance tasks (Scialfa et al., 1988). Thus far, there are very few studies of age-differences in DVA under various levels of illumination. In one recent study, differences in SVA, DVA, binocular, horizontal, and peripheral visual fields, and several driving-related performance variables were compared between 70 year old drivers and age-matched control non-drivers. Both SVA and DVA were measured at three luminance levels: 1) daytime (55 foot-candles), 2) twilight (30 ft-cd), and 3) nighttime (10 ft-cd). In both SVA and DVA, targets were presented at 14 inches, and in DVA, these targets were presented at velocities of 1.5 feet/second (an estimated pedestrian walking speed), in the horizontal axis or meridian. According to multivariate evaluations, dynamic acuity, peripheral visual fields, reaction time, and non-dominant hand grip were significant predictors of driving frequency, and distinguishing variables between 70 year old drivers and age-matched non-drivers. According to univariate comparisons of SVA and DVA, non-drivers had SVA worse than 20/40 in either eye. Prominent differences between SVA and DVA became most apparent at twilight and nighttime levels of illumination between drivers and non-drivers (Retchin, et al., 1988). Thus far, remarkably few studies have compared SVA and DVA in terms of possible causes for acuity loss in DVA. Generally, DVA has been restricted to measurements of targets moving at a constant angular velocity and at a constant distance from the observer. DVA is most often expressed as threshold target size in terms of min. of arc of the gap in a Landolt ring that can be resolved at a given target angular velocity. Two earlier but notable studies examined DVA thresholds for moving Landolt ring targets at the fovea and peripheral retina (Brown, 1972a), as well as in relation to peripheral acuity for those moving targets (Brown, 1972b). At the fovea, it was observed that the relation between thresholds (min. of arc) and target image angular velocity was linear, at velocities from 0 to 50 deg/sec and eccentricities from 0 to 10 degrees. In the parafovea, the relation was more complex and acuity was better for slowly moving targets than for static targets. It was concluded that different neural mechanisms may mediate spatial resolution of moving targets in the parafovea. As yet, only age-differences in foveal cone density, but not in peripheral cones and rods have been evaluated in human, and nonhuman subjects (Ordy, et al., 1991).

In a recent study, correlations were compared among static acuity (SVA), spatial contrast sensitivity (SVS), and dynamic spatial contrast sensitivity (DVS) in 49 subjects ranging from 7 to 61 years of age. DVS was measured at 1.5 to 16.5 c/deg, at 18 inches, with luminance of the targets at 20.42 ft-L, and with a circular motion of the target. The subjects were required to discriminate grating orientation in DVS. Contrary to some previous findings, there was high correlation between SVS and DVS, even at higher spatial frequencies. SVA was poorly correlated with DVS. In terms of the effects of age, both SVS and DVS declined with age, with DVS being more affected by age. SVS and DVS were also significantly correlated (Scialfa, et al., 1988). Although these recent studies of age-differences in relationships among measures of static and dynamic contrast sensitivity appear of some theoretical interest for studies of possible age-differences in optical, retinal, and geniculostriate mechanisms that may be involved in spatial contrast sensitivity, visual tracking, and oculomotor motility, they also introduce a whole new level of complexity to assessments of dynamic visual acuity, contrast sensitivity, and driving. Adequate control of test conditions involving increasing variance with age, lower luminance, and increasing target velocity, as well as administration of these tests, analyses of data, and interpretation of results, become much more difficult (Retchin, et al., 1988; Long and Garvey, 1988; Scialfa, et al., 1988).

AGE DIFFERENCES IN THE EYE, RETINA, AND GENICULOSTRIATE SYSTEM RELEVANT FOR ASSESSMENT OF VISUAL FUNCTIONS INVOLVED IN DRIVING

Visual acuity depends upon optical and neural mechanisms extending from the eye to the visual pathways and centers of the brain (Ordy, et al., 1991). Ocular components influencing acuity include cornea, lens, pupil, ocular media, and possibly the retinal pigment epithelium (Marmor, 1982). Neural factors in acuity include foveal cone and ganglion cell density, their ratio and eccentricity from the foveola, a magnification factor in the geniculostriate system, and ultimately, receptive fields of neurons for spatial orientation, contrast, and frequency in the foveal striate cortex (Rodieck, 1979; Van Essen, 1979; Wässle, et al., 1987; Schein, 1988). Some progress has been made in vision research concerning the respective roles of optical and neural factors in normal postnatal development of visual acuity (Teller and Movshon, 1986; Jacobs and Blakemore, 1988). Much less is known about the optical and neural factors in visual acuity declines from maturity to senescence (Ordy, et al., 1991). Because the role of the eye in visual acuity is more accessible to psychophysical, electrophysiological, and ophthalmological evaluations, considerably more emphasis has thus far been placed on optical and funduscopically observable retinal factors in acuity declines of the elderly, compared to possible age-related changes, such as cell loss in the retina, and geniculostriate system (Ordy, et al., 1991).

Because the cornea, lens, pupil, and ocular media determine the retinal image in acuity, age-dependent changes in the lens, pupil, and ocular media are known to result in reduced illumination, increased light scatter, and a blurring of the retinal image, producing reduced acuity, particularly for targets of low illumination, contrast, and high spatial frequency (Carter, 1982; Weale, 1982). Decreased pupillary diameter, or miosis, and opacification of the lens, (cataract formation), and yellowing with increasing age, are closely associated with changes in such diverse visual functions as acuity, range of accommodation, presbyopia, contrast sensitivity, field of vision, glare sensitivity, and color vision (Carter, 1982; Spector, 1982; Weale, 1982).

Studies of age-related changes in neural mechanisms of acuity from maturity to senescence have thus far focused on foveal cone density, ganglion cell loss, or degeneration, and loss of neurons in the geniculostriate system (Ordy, et al., 1991). A tentative hypothesis in neuroscience of vision has proposed that significant correlations can be made between such basic visual function as acuity, foveal cone density (Williams, 1986), and the spatial densities of foveal representations in the striate cortex (Schein, 1988). Correlations have been reported among loss of photoreceptors, lipofuscin accumulation in retinal pigment epithelium (RPE), loss of melanin in RPE, and macular degeneration in the human retina with age (Dorey, et al., 1989). In preliminary quantitative morphometric estimates of age-differences in cone density of the "rod-free" foveola of the fovea, the number of cones was estimated at 76,282 in a 37 year old, and at 48,804 cones/mm^2 by 72 years, in this central-most 250 μm foveola (Youdelis and Hendrickson, 1986). In addition to the age-related changes in RPE pigments, photoreceptors, their outer segments, or discs, and retinal ganglion cells, that may be causally associated with declines of acuity, it also seems important to note that an estimated loss of 54% of neurons has been reported from age 20 to 87 years in the foveal projection area of the striate cortex in man (Devaney and Johnson, 1980). Declines in visual acuity and foveal cone density have also been reported in the retina of the aged rhesus monkey (Ordy, et al., 1980). Although considerable information is becoming available concerning the physiological and anatomical bases of visual acuity at the level of the human (Hirsch and Curcic, 1989) and nonhuman primate retina (Ordy, et al., 1980), as well as the nonhuman primate geniculostriate system (Wässle, et al., 1987; Schein, 1988), considerably less is known about the molecular and neural changes that may be involved in age-related declines of visual acuity in the human visual system (Ordy, et al., 1991).

VISION, AGING, DRIVING: OCULAR ANOMALIES, PATHOLOGIES, OPHTHALMOLOGICAL PROBLEMS OF THE ELDERLY

As a starting point, it is apparent that a major challenge for basic and clinical research on vision, aging, and driving is the development of criteria for the distinction between the effects of normal aging and those of ocular pathology. Age-related declines in visual acuity and other function, without ocular pathology are often encountered in non-clinical and in clinical settings. It should also be apparent that the basic and applied studies of visual acuity and psychomotor performance essential for driving by the elderly have thus far been reviewed in the previous sections without reference to specific ocular pathologies. For example, age-related gradual changes in the pupil, lens, ocular media, photoreceptors, and in RPE are likely to result in changes in such diverse visual functions as acuity, contrast and glare sensitivity, field of vision, adaptation, and accommodation (Carter, 1982). Visual impairments with increasing age have become more prominent problems for such disciplines as psychology, optometry, ophthalmology, and gerontology to a considerable extent because of the increasing proportion of elderly. Visual impairments with increasing age, however, become manifest through both normal and pathological processes. In terms of the roles of optometrists and ophthalmologists in assessments of visual acuity impairments and driving, in younger and some middle-aged drivers, refractive errors may represent the more common problems, and in most cases, correction of visual impairments with glasses is readily possible. In elderly drivers, declines in acuity may be due to ocular pathologies, which for the most part are not amenable to correction by glasses. Until recently, ophthalmological studies of vision and aging have focused mainly on visual impairments, including acuity, more directly related to such disease entities as cataracts, glaucoma, and more recently to age-related macular degeneration (AMD) (Greenberg and Branch, 1982; Lovie-Kitchin and Bowman, 1985). Earlier reviews of methodological issues concerning the incidence and prevalence of visual impairments and ocular pathologies in the elderly have recognized limitations in clinical "case" reports, and other sources of inadequate information or data that may result in "under-reporting" of these visual impairments and ocular pathologies in the elderly (Greenberg and Branch, 1982). Despite methodological sampling limitations, cataracts, glaucoma, and retinal disorders other than diabetic retinopathy, appear to be six to eight times more prevalent among individuals age 65 years or more (Greenberg and Branch, 1982).

Perhaps in recognition of the increasing proportion of elderly drivers with visual impairments and ocular pathologies, a number of ophthalmological surveys have been undertaken of visual functions, driving safety, and the elderly (Harms, et al., 1984; Graca, 1986; Harms and Dietz, 1987; Keltner and Johnson, 1987). Despite the methodological limitations of even extensive ophthalmological surveys, some startling findings have been reported concerning the prevalence of inadequate vision, incidence of ocular pathologies, and the lack of awareness, or even disregard of visual problems, with adverse implications for driving, particularly during nighttime conditions (Harms, et al., 1984). In young, and some middle-aged drivers, refractive errors appeared to be the primary cause of inadequate vision, and in most cases correction with glasses was possible. In elderly drivers, the loss of acuity was mostly associated with opacities of the lens. Surprisingly, two-thirds of the drivers with "poor" acuity considered their vision to be adequate for driving, with younger drivers being more aware than older drivers of their inadequate vision. Only 30% of visually impaired drivers stopped driving at night voluntarily. As many as two-thirds of the drivers that considered their acuity to be impaired continued to drive a motor vehicle at night (Harms, et al., 1984). Recognizing that the physician may be in a critical position for monitoring visually impaired and unsafe elderly drivers, ophthalmological surveys have suggested a more active role for physicians in detection of unsafe drivers, and for making recommendations concerning regulations on restrictions and cancellation of a driver's license. However, questions concerning confidential patient/physician relationships, and legislation regarding physician liability need to be clarified (Graca, 1986). In ophthalmological surveys by the Department of Motor Vehicles in all 50 states (see Table 1), concerning visual function, driving safety, and the elderly, many common visual problems were evaluated, and specific recommendations were made for identifying unsafe drivers, setting criteria for restrictions on a driver's license, and balancing the needs of individual drivers with traffic safety considerations (Keltner and Johnson, 1987). Recommended criteria for a non-restricted driver's license included: 1) SVA of 20/40 in either or both eyes, with or without corrective lenses, and 2) an uninterrupted visual field of 140°, determined with a 1° test object. In addition to the specific standards of requirements for a restricted and non-restricted driver's license, the following recommendations were also made concerning visual functions, driving safety, and the elderly: 1) drivers over 65 years of age should have SVA and visual fields tested every 1 to 2 years, 2) regulations need to be formulated enabling concerned family members and physicians to report a questionable driver for an examination for a driver's license, and 3) drivers of any age with convictions and multiple accidents should be required to have SVA and visual fields evaluated (Keltner and Johnson, 1987). It was also recommended that because elderly drivers do not manifest age-related visual and motor performance declines at the same rate, they should be evaluated in terms of functional standards rather than according to some arbitrary criterion of chronological age.

VISION, AGING, DRIVING IN CONTEXT OF OTHER HUMAN FACTORS

Although primary emphasis in this review has been on age-related impairments in visual acuity, peripheral fields, contrast sensitivity, and the increasing incidence of ocular pathologies with direct relevance to driving, increasingly, measurements of visual acuity are being made in relation to other visual functions, including divided visual attention in driving, impaired cognitive capacity, decreased choice reaction time and other specific psychomotor skills essential for driving (Retchin, et al., 1988). Even some earlier studies recognized that age-related declines occurred not only in visual acuity and visual field size, but also in visual search, and psychomotor skills, particularly in the elderly driver (Paneck, et al., 1977). More recently, through engineering improvements, automobiles have become faster and they have more controls and options so that decisions must be made at a more rapid rate. Interstate highways have been built for faster traffic. Significant age-differences have been reported in divided attention involving a visual tracking task, and self-paced visual choice-reaction time in a simulated driving task (Ponds, et al., 1988).

It seems obvious that problems of vision and driving include problems of speed, reaction time, stationary and moving visual target characteristics, as well as vehicle handling characteristics and driving strategies (Summala, 1981; Probst, et al., 1984; Godthelp and Käppler, 1988). It is also apparent that driving safety, particularly in nighttime driving by the elderly will become of increasing concern for development of multivariate tests of vision and motor performance, increased awareness and education, enforcement, and engineering aspects of vehicle safety. For example, there is considerable need for development of a multivariate test battery that would include static and dynamic acuity, spatial contrast sensitivity, measured under high and low levels of illumination, and glare sensitivity, which might be useful for screening elderly drivers whose vision declines significantly under low levels of illumination (Sturgis and Osgood, 1982; Sturr, et al., 1990). There is a special need in studies of vision, aging and driving for interaction and collaboration among vision scientists, clinicians, and human factors' engineers. Age-related changes in other sensory functions, learning, and motivation of elderly drivers are also receiving attention, particularly in relation to traffic safety and accidents (Winter, 1985).

VISION, DRIVING, SAFETY: TRAFFIC VIOLATIONS AND ACCIDENT RATES IN THE ELDERLY

As indicated in this review, there is evidence that visual functions and motor skills needed for safe driving, particularly at night, begin to decline with increasing age to 65, and more noticeably after the age of 75. Although the elderly may drive 30% to 50% fewer miles than younger drivers (Graca, 1986), they appear to have higher accident rates than younger drivers excluding those 25 years and younger (Klamm, 1985). Whereas many studies and surveys have examined correlations among age, traffic violation, and accidents, thus far few studies have attempted to correlate age-related impairments in specific visual function, such as acuity, and specific motor skills, with specific traffic violations and accident rates in the elderly (Keltner and Johnson, 1987). Major sources for statistics on visual

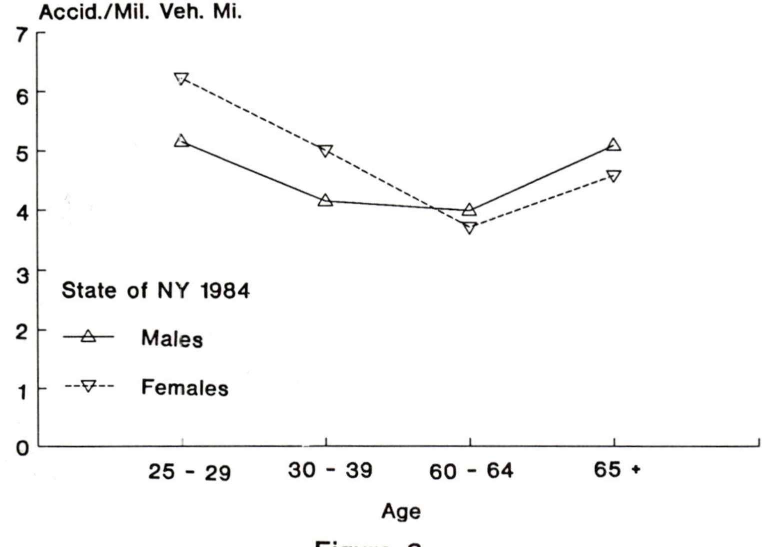

Figure 3

Age and Sex Differences in Accident Rates

standards, driving characteristics, traffic violation, and accident rates include: 1) US Department of Transportation, 2) individual state departments of motor vehicles or transportation, 3) auto insurance companies or agencies (Klamm, 1985), and 4) consultants to the Transportation Research Board of the AAA Foundation for Traffic Safety (Solomon, 1985).

In a major survey of visual functions and driving safety among the elderly, most states reported that the elderly were involved in a relatively small percentage of the total number of accidents (Keltner and Johnson, 1987). In the same states, it was not taken into consideration that elderly drivers tend to drive considerably fewer miles compared to younger age groups. Several states have focused on accident rates per 100,000 miles driven as a function of age and sex. Interestingly, both older and younger drivers had higher accident rates per 100,000 miles driven compared to middle aged groups. This pattern of accident rates by age and sex across the years from 25 to 65 and above is illustrated in Figure 3, for the state of New York, for the year 1984 (see Figure 3).

The California Department of Transportation has conducted extensive surveys on the vision screening of driver's license applicants, including the elderly. The pattern of accident rates as functions of age and sex for 1982 was very similar to the pattern observable in the state of New York (see Figure 3). In addition to the descriptive statistics of accident rates as function of age and sex, data were also presented on the types of traffic violations that occurred as a function of age in California, in 1982. As expected, speed was a major variable in violations per 100,000 miles only for the 20 to 25 year old group, whereas violations that occurred most often in the elderly included failure to yield right-of-way violations involving turns, stop signs, and oncoming traffic. Figure 4 illustrates traffic violations by type and age for the state of California for 1982 (Keltner and Johnson, 1987, p. 1185, Figure 2, modified).

Based on assumptions that experience, awareness, learning, attitudes, and motivation can compensate for some of the age-related sensory,

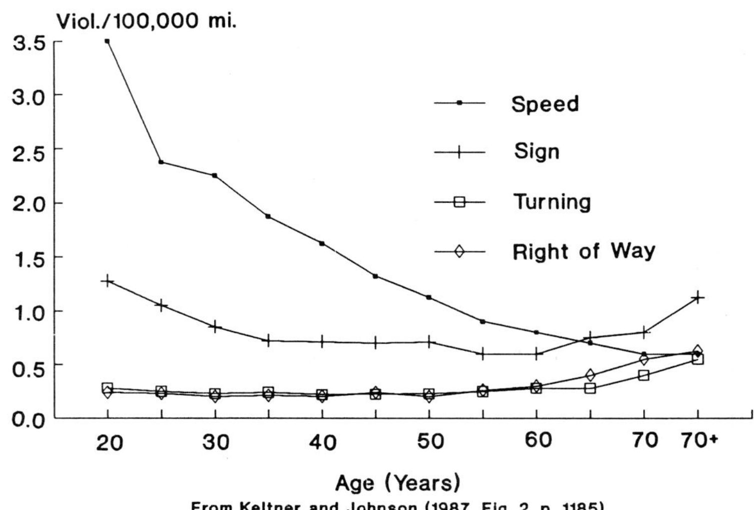

From Keltner and Johnson (1987, Fig. 2, p. 1185).

Figure 4

Traffic Violations by Type and Age

cognitive, and performance declines, a national group, the American Association of Retired Persons (AARP), has developed an extensive nationwide course called "55 Alive - Mature Driving Course", to help elderly with their driving skills and prevent accidents (Milone, 1985). Possible evidence for the success of the program may be the reduction of insurance premiums in several states (Charman, 1985).

VISION AND ELDERLY DRIVING PERFORMANCE: URBAN, HIGHWAY CONDITIONS, TRAFFIC SIGNS, VEHICLE DESIGNS, POSSIBLE MODIFICATIONS

It seems obvious that urban, or highway conditions, traffic signs and control devices, vehicle characteristics, and traffic regulations were designed primarily for adult drivers with optimum, or at least adequate visual functions, cognitive ability, and motor skills essential for safe driving. As long as the proportion of elderly drivers of the total population of drivers remained negligible, major concerns for making compensatory modifications in urban or highway conditions, traffic signs, or vehicle characteristics to improve driving conditions and safety for the elderly driver did not appear as major challenges. In addition to helping identify drivers with significant visual impairments, ocular pathologies, and psychomotor performance problems, as outlined in this review, it should also be possible to make driving easier and safer for many elderly drivers by also identifying possible modifications in roadway conditions, traffic signs and control devices, and vehicle designs (Keltner and Johnson, 1987).

Perhaps, due to increasing awareness of the needs and problems of older drivers by the general population of drivers, as well as the elderly drivers themselves (Malfetti, 1985), and due to greater awareness among visual scientists and human factors engineers, there is a modest increase in the number of basic, clinical, and applied research studies on vision, aging, and driving. There are also survey results and recommendations concerning elderly driver characteristics, vehicle designs, and highway conditions (Allen, 1985). Human factors studies have focused on such relevant problems as: the effects of target wavelength on dynamic visual acuity under photopic and scotopic conditions (Long and Garvey, 1988), effects of instrument panel luminance and chromaticity on reading performance (Imbeau, et al., 1989), movement time and brake pedal placement (Morrison, et al., 1986), and development of time-related measures to describe driving strategies (Godthelp, et al., 1984). As yet, these type of human factors studies have only begun to focus on important and interacting environmental, vehicular, and driver characteristics in terms of improving driving safety in the elderly.

There are numerous surveys of elderly driver's characteristics, safety aspects of vehicle designs, and possible modifications of highway driving conditions for improved safety (Malfetti, 1985). Surveys have identified some of the following needs and problems of elderly drivers, with specific recommendations for remedial implementation: 1) impairments of vision in elderly drivers, with specific implications for vehicle and highway designs and for more frequent driver testing (Allen, 1985), 2) learning and motivational characteristics of elderly drivers pertaining to improvements of traffic safety (Winter, 1985), 3) education concerning possible adverse effects of medication and drugs used by elderly drivers (Malfetti, 1985), and 4) special auto insurance needs and problems of elderly drivers (Klamm, 1985). Although many authors of these surveys recognize the need for more basic, clinical, and applied research on vision, aging, and driving, they also suggest that the elderly drivers themselves may represent a good source of information, and starting point for identifying their problems, and helping with improvements.

RESEARCH OPPORTUNITIES FOR IDENTIFYING SOME MAJOR PROBLEMS OF VISION, AGING, AND DRIVING WITH IMPLICATIONS FOR ELDERLY DRIVERS

There are at least three major inter-related demographic and biological issues in vision, aging, and driving that provide a challenge for visual scientists, human factors engineers, policy planners, and regulatory agencies: 1) through socioeconomic, nutritional, and medical advances, as well as secular trends, the median lifespan, health, and functional status of the elderly seems to be improving, 2) there is credible evidence, presented in this review, that visual functions and psychomotor skills essential for driving may begin to decline from age 55 to 65, and then more prominently after age 75, and 3) that not only basic univariate science studies, but multidisciplinary assessments of visual and psychomotor functions are needed with reference to interacting driver characteristic, environmental conditions, and vehicular factors. Attempts were made in this review to describe current visual driving standards, identify basic, clinical, and applied research topics with direct implications for vision driving standards for the elderly. Attempts were also made to highlight some research opportunities for optimizing visual functions and driving performance by the increasing proportion of elderly drivers. Some specific research opportunities were identified in the following areas: 1) scope of the vision, aging, and driving problem in terms of demographic changes in proportion of elderly, and elderly drivers in total population, 2) variability among states in general vision standards or requirements for driving in a changing "mixed-age" driving population, 3) the presumed "face" validity of a 20/40 static visual acuity (SVA) standard, tested at photopic illumination, as a predictor of the "dynamic visual acuity (DVA) criterion variable" of driving under photopic, mesopic, and scotopic illumination, particularly in relation to traffic safety, 4) research evidence for recommending modification of the 20/40 SVA acuity standard for low levels of illumination for a nighttime acuity standard, 5) basic and applied innovative studies of age-differences in relationships between SVA and DVA under photopic, mesopic, and scotopic illumination, 6) generation of data for possible development of a multivariate test battery including SVA, DVA, and spatial contrast sensitivity measures under photopic, mesopic, and scotopic levels of illumination, 7) correlational studies of age-differences in acuity, visual fields, and spatial contrast sensitivity, with optical variables, foveal cone/rod density, and other neural mechanisms of acuity and spatial contrast sensitivity, 8) correlational studies of age-differences in such visual impairments as acuity and peripheral fields, with ocular anomalies and pathologies, 9) greater emphasis of multidisciplinary, or multivariate assessments of vision, aging, and driving in context of motor skills and other human factors, 10) shift in emphasis from descriptive statistics to more appropriate statistical evaluations of types and frequencies of traffic violations among elderly drivers, and 11) feasibility of human factors and engineering studies of possible modifications of urban, highway conditions, traffic signs, control devices, vehicle designs, and traffic regulations to improve safety in elderly drivers. It seems clear that there is a considerable need for basic, clinical, and applied research to accumulate a credible database. There is also a need for a continuous communication or "dialogue" among scientists, human factors engineers, policy planners, and regulatory agencies, as well as the elderly drivers themselves concerning emerging problems in vision, aging, and driving.

REFERENCES

Adams, A.J., Wong, L.S., Wong, L., and Gould, B., 1988, Visual acuity changes with age: Some new perspectives, Am. J. Optom. Physiol. Opt., 65:403-406.

Allen, M.J., 1985, Vision of the older driver: Implications for vehicle and highway design and for driver testing, in: Drivers 55+, Needs and problems of older drivers: Survey results and recommendations, Proceedings of the older driver colloquium, J.L. Malfetti, ed., AAA Foundation for Traffic Safety, Falls Church, Va., pp. 1-7.

Birren, J.E., and Williams, M.V., 1982, A perspective of aging and visual function, in: "Aging and human visual function," R. Sekuler, D. Kline, and K. Dismukes, eds., Alan R. Liss, Inc., New York, NY, pp. 7-19.

Brown, B., 1972a, Resolution thresholds for moving targets at the fovea and in the peripheral retina, Vision Res., 12:293-304.

Brown, B., 1972b, Dynamic visual acuity, eye movements and peripheral acuity for moving targets, Vision Res., 12:305-321.

Burg, A., 1966, Visual acuity as measured by dynamic and static tests, J. Appl. Psych., 50:460-466.

Burg, A., 1968, The relationship between vision test scores and driving record: Additional findings, report 68-27, UCLA Dept. of Engineering, Los Angeles, CA.

Carter, J.H., 1982, The effects of aging upon selected visual functions: Color vision, glare sensitivity, field of vision, and accommodation, in: "Aging and human visual function," R. Sekuler, D. Kline, and K. Dismukes, eds., Alan R. Liss, Inc., New York, NY, pp. 121-130.

Charman, W.N., 1985, Visual standards for driving, Ophthalmic. Physiol. Opt., 5:211-220.

Devaney, K.O., and Johnson, H.A., 1980, Neuron loss in the aging visual cortex of man, J. Gerontol., 35:836:841.

Dorey, C.K., Wu, G., Eberstein, D., Farsd, A., and Weiter, J.J., 1989, Cell loss in the aging retina, Invest. Ophthalmol. Vis. Sci., 30:1691-1699.

Evans, D.W., and Ginsburg, A.P., 1985, Contrast sensitivity predicts age-related differences in highway-sign discriminability, Human Factors, 27:637-642.

Federal Highway Administration, 1986, Driver's license administration requirements and fees, U.S. Dept. of Transportation, Washington, D.C.

Fonda, G., 1986, Suggested vision standards for drivers in the United States with vision ranging from 20/175 (6/52) to 20/50 (6/15), Ann. Ophthalmol., 18:76-79.

Godthelp, H., and Kappler, W.D., 1988, Effects of vehicle handling characteristics on driving strategy, Human Factors, 30: 219-229.

Godthelp, H., Milgram, P., and Blaauw, G., 1984, The development of a time-related measure to describe driving strategy, Human Factors, 26:257-268.

Graca, J.L., 1986, Driving and aging, Clin. Geriatr. Med., 2:577-589.

Greenberg, D.A., and Branch, L.G., 1982, A review of methodological issues concerning incidence and prevalence data of visual deterioration in elders, in: "Aging and human visual function," R. Sekuler, D. Kline, and K. Dismukes, eds., Alan R. Liss, Inc., New York, NY, pp. 279-296.

Harms, H., and Dietz, K., 1987, Causes of inadequate vision in drivers and their significance in traffic, Klin. Monatsbl. Augenheilkd., 190:3-7.

Harms, H., Kroner, B., and Dannheim, R., 1984, Ophthalmological experiences with automobile drivers with inadequate vision, Klin. Monatsbl. Augenheilkd., 185:77-85.

Hirsch, J., and Curcio, C.A., 1989, The spatial resolution capacity of the human fovea, Vision Res., 29:1095-1101.

Imbeau, D., Wierwille, W.W., Wolf, L.D., and Chun, G.A., 1989, Effects of instrument panel luminance and chromaticity on reading performance and preference in simulated driving, Human Factors, 31:147-160.

Jacobs, D.S., and Blakemore, C., 1988, Factors limiting the postnatal development of visual acuity in the monkey, Vision Res., 28:947-958.

Keltner, J.L., and Johnson, C.A., 1987, Visual function, driving safety, and the elderly, Ophthalmol., 94:1180-1188.

Klamm, E.R., 1985, Auto insurance: Needs and problems of drivers 55 and over, in: Drivers 55+, Needs and problems of older drivers: Survey results and recommendations, Proceedings of the older driver colloquium, J.L. Malfetti, ed., AAA Foundation for Traffic Safety, Falls Church, Va., pp. 21-31.

Kosnik, W., Winslow, L., Kline, D., Rasinski, K., and Sekuler, R., 1988, Visual changes in daily life throughout adulthood, J. Gerontol., 43:63-70.

Long, G.M., and Garvey, P.M., 1988, The effects of target wavelength on dynamic visual acuity under photopic and scotopic viewing, Human Factors, 30:3-13.

Lovie-Kitchin, J.E., and Bowman, K.J., 1985, "Senile macular degeneration," Butterworth Publishers, Boston.

Malfetti, J.L., 1985, Drivers 55+, Needs and problems of older drivers: Survey results and recommendations, Proceedings of the older driver colloquium, AAA Foundation for Traffic Safety, Falls Church, Va.

Marmor, M.F., 1982, Aging and the retina, in: "Aging and human visual function," R. Sekuler, D. Kline, and K. Dismukes, eds., Alan R. Liss, Inc., New York, NY, pp. 59-78.

Milone, A.M., 1985, Training and retraining the older driver, in: Drivers 55+, Needs and problems of older drivers: Survey results and recommendations, Proceedings of the older driver colloquium, J.L. Malfetti, ed., AAA Foundation for Traffic Safety, Falls Church, Va., pp. 38-44.

Morrison, R.W., Swope, J.G., and Halcomb, C.G., 1986, Movement time and brake pedal placement, Human Factors, 28:241-246.

Ordy, J.M., Brizzee, K.R., and Hansche, J., 1980, Visual acuity and foveal cone density in the retina of the aged rhesus monkey, Neurobiol. Aging, 1:133-140.

Ordy, J.M., Wengenack, T.M., and Dunlap, W.P., 1991, Visual acuity, aging, and environmental interactions: A neuroscience perspective, in: "The effects of age and environment on vision," D. Armstrong, M.F. Marmor, J.M. Ordy, eds., Plenum Press.

Owsley, C., Sekuler, R., and Siemsen, D., 1983, Contrast sensitivity throughout adulthood, Vision Res., 23:689-699.

Panek, P.E., Barrett, G.V., Sterns, H.L., and Alexander, R.A., 1977, A review of age changes in perceptual information ability with regard to driving, Exp. Aging Res., 3:387-449.

Pitts, D.G., 1982, The effects of aging on selected visual functions: Dark adaptation, visual acuity, stereopsis, and brightness contrast, in: "Aging and human visual function," R. Sekuler, D. Kline, and K. Dismukes, eds., Alan R. Liss, Inc., New York, NY, pp. 131-159.

Ponds, R.W.H.M., Brouwer, W.H., and van Wolffelaar, P.C., 1988, Age differences in divided attention in a simulated driving task, J. Gerontol., 43:P151-156.

Probst, T., Krafczyk, S., Brandt, T., and Wist, E.R., 1984, Interaction between perceived self-motion and object-motion impairs vehicle guidance, Science, 225:536-538.

Reading, V.M., 1972, Visual resolution as measured by dynamic and static tests, Pflugers Arch., 333:17-26.

Retchin, S.M., Cox, J., Fox., and Irwin, L., 1988, Performance-based measurements among elderly drivers and nondrivers, J. Am. Ger. Soc., 36:813-819.

Richards, O.W., 1977, Effects of luminance and contrast on visual acuity, ages 16 to 90 years, Am. J. Optom. Phys. Opt., 54:178-184.

Rodieck, R.W., 1979, Visual pathways, Ann. Rev. Neurosci., 2:193-225.

Rosenbloom, S., 1988, The mobility needs of the elderly, in: Transportation in an aging society, vol. 2, Transportation Research Board, National Research Council, Washington, D.C.

Schein, S.J., 1988, Anatomy of macaque fovea and spatial densities of neurons in foveal representation, J. Comp. Neurol., 269:479-505.

Scialfa, C.T., Garvey, P.M., Gish, K.W., Deering, L.M., Leibowitz, H.W., and Goebel, C.C., 1988, Relationships among measures of static and dynamic visual sensitivity, Human Factors, 30:677-687.

Sekuler, R., Hutman, L.P., and Owsley, C., 1980, Human aging and spatial vision, Science, 209:1255-1256.

Sekuler, R., and Owsley, C., 1982, The spatial vision of older humans, in: "Aging and human visual function," R. Sekuler, D. Kline, and K. Dismukes, eds., Alan R. Liss, Inc., New York, NY, pp. 185-202.

Shinar, D., and Drory, A., 1983, Sign registration in daytime and nighttime driving, Human Factors, 25:117-122.

Siegel, J.S., 1980, Recent and prospective trends for the elderly population and some implications for health care, in: "Second conference on the epidemiology of aging," S.G. Haynes and M. Feinleib, eds., U.S. Dept. of Health and Human Services, Washington, D.C., NIH Publication No. 80969.

Sivak, M., Olson, P.L., and Pastalan, L.A., 1981, Effects of driver's age on nighttime legibility of highway signs, Human Factors, 23:59-64.

Sloane, M.E., Owsley, C., and Alvarez, S.L., 1988, Aging, senile miosis and spatial contrast sensitivity at low luminance, Vision Res., 28:1235-1246.

Solomon, D., 1985, The older driver and highway design, in: Drivers 55+, Needs and problems of older drivers: Survey results and recommendations, Proceedings of the older driver colloquium, J.L. Malfetti, ed., AAA Foundation for Traffic Safety, Falls Church, Va., pp. 55-62.

Spector, A., 1982, Aging of the lens and cataract formation, in: "Aging and human visual function," R. Sekuler, D. Kline, and K. Dismukes, eds., Alan R. Liss, Inc., New York, NY, pp. 27-43.

Storandt, M., 1982, Concepts and methodological issues in the study of aging, in: "Aging and human visual function," R. Sekuler, D. Kline, and K. Dismukes, eds., Alan R. Liss, Inc., New York, NY, pp. 269-278.

Sturgis, S.P., and Osgood, D.J., 1982, Effects of glare and background luminance on visual acuity and contrast sensitivity: Implications for driver night vision testing, Human Factors, 24:347-360.

Sturr, J.F., Kline, G.E., and Taub, H.A., 1990, Performance of young and older drivers on a static acuity test under photopic and mesopic luminance conditions, Human Factors, 32:1-8.

Summala, H., 1981, Driver/vehicle steering response latencies, Human Factors, 23:683-692.

Teller, D.Y., and Movshon, J.A., 1986, Visual development, Vision Res., 26:1483-1506.

Van Essen, D.C., 1979, Visual areas of the mammalian cerebral cortex, Ann. Rev. Neurosci., 2:227-263.

Wässle, H., Grunert, U., Rohrenbeck, J., and Boycott, B.B., 1987, Retinal ganglion cell density and cortical magnification factor in the primate, Vision Res., 30:1897-1911.

Weale, R.A., 1982, Senile ocular changes, cell death, and vision, in: "Aging and human visual function," R. Sekuler, D. Kline, and K. Dismukes, eds., Alan R. Liss, Inc., New York, NY, pp. 161-171.

Williams, D.R., 1986, Seeing through the photoreceptor mosaic, Trends in Neurosci., 9:193-198.

Winter, D.J., 1985, Learning and motivational characteristics of older people pertaining to traffic safety, in: Drivers 55+, Needs and problems of older drivers: Survey results and recommendations, Proceedings of the older driver colloquium, J.L. Malfetti, ed., AAA Foundation for Traffic Safety, Falls Church, Va., pp. 77-86.

Youdelis, C., and Hendrickson, A., 1986, A qualitative and quantitative analysis of the human fovea during development, Vision Res., 26:847-855.

MATURATION AND AGING ARE ASSOCIATED WITH DECREMENTS IN UBIQUITIN-PROTEIN CONJUGATION CAPABILITY: PLAUSIBLE INVOLVEMENT OF IMPAIRED PROTEOLYTIC CAPABILITIES IN CATARACT FORMATION

Allen Taylor and Jessica Jahngen-Hodge

Laboratory for Nutrition and Vision Research
USDA Human Nutrition Research Center On Aging at
Tufts University
711 Washington Street
Boston, MA 02111

INTRODUCTION

The eye lens is composed of about 40% (by weight) proteins, the remainder being water. Protein turnover in the lens occurs at only marginal rates (1), particularly in the older lens tissue (2). Older tissue is found in the lens nucleus, whereas younger, less developed and newly elaborated tissue is found in the outer lens tissue layers, also called cortex and epithelium as illustrated in Figure 1. Due to this segregation of tissue the lens is a useful tissue in which to study aspects of protein metabolism as it is affected by development and aging.

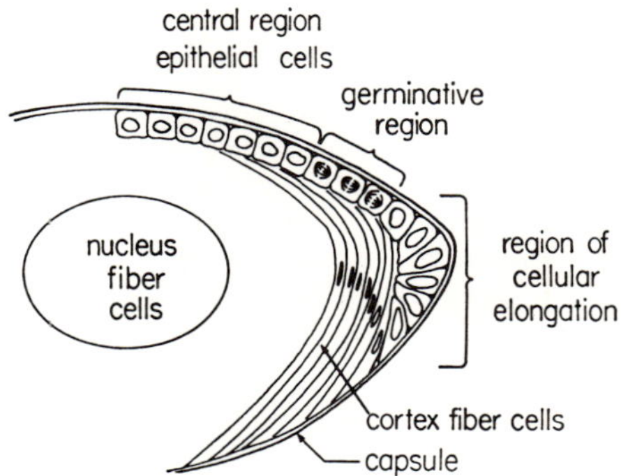

Figure 1. Diagram of the lens of the adult vertebrate. The lens is surrounded by an external noncellular capsule. Beneath the capsule are the lens epithelial cells. In the equatorial zone the epithelial cells begin to elongate into fiber cells. Elongation is concomitant with enhanced production of the major lens gene products, or crystallins. Fiber cells that are newly laid down constitute the cortex region; the fiber cells formed during the early growth period of the lens compose the nucleus region of the adult lens. This figure was adapted with permission from Dr. Papaconstantinou (3).

Upon aging, the lens is exposed to light and oxygen. It is estimated that over 90% of major lens proteins found in aged and cataractous lens are postsynthetically modified (4). Such damage is associated with the eventual aggregation and precipitation of proteins in cataracts from the clear lens milieu (5).

In most cell systems studied damaged proteins are more rapidly removed than are native proteins. This editing is apparently accomplished by a group of proteolytic enzymes which act in concert with a molecular machinery for identifying and labeling damaged proteins for degradation (6.7). The system for identifying and labeling damaged proteins which has been recently described involves conjugation of ubiquitin, a 76 amino acid polypeptide, to putative substrate proteins (Fig.2). In this paper we show that the lens also contains the elements of the ubiquitin system necessary to target Proteins for degradation. However, we present date which indicates that, like the activities of many lens proteases (for review, see ref. 2 and 8-11), these ubiquitin-dependent activities are attenuated during development and aging (12.13). This might explain why only a minor extent of proteolysis is detected on a whole lens basis. We also discuss the possibilities that attenuated proteolytic capabilities may be related to the accumulation of damaged proteins in the aged lens and that unchecked accumulation of such proteins may be associated with cataract formation (2).

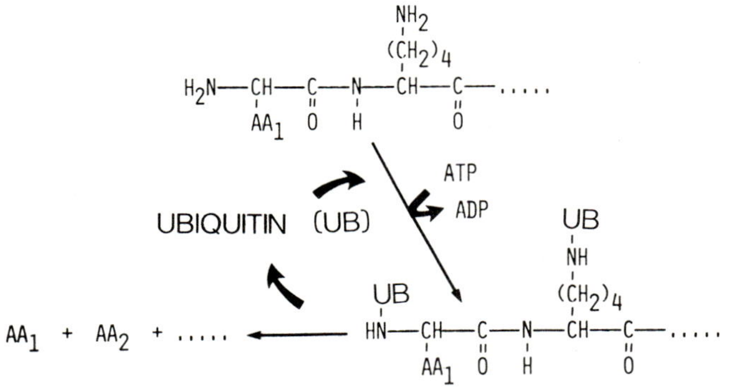

NH_2
$(CH_2)_4$
$H_2N—CH—C—N—CH—C—.....$
AA_1 O H O
ATP
ADP
UBIQUITIN (UB)
UB
NH
$(CH_2)_4$
UB
$AA_1 + AA_2 + ...$ ← $HN—CH—C—N—CH—C—.....$
AA_1 O H O

Figure 2. Schematic of protein degradation via the cytoplasmic ubiquitin/ATP dependent pathway. Damaged or rapidly turned over proteins have ubiquitin conjugated to their epsilon and possibly alpha amino termini. The ubiquitin-protein conjugates are processed for degradation such that many of the ubiquitin moieties are recycled while peptides are liberated from the lens protein portion. The polypeptides are hydrolyzed by endo- and exopeptidases. This figure was adapted from Hershko (6).

METHODS

Reagents for ubiquitination assays and preparation of culture rabbit lens epithelial cell supernatants and ^{125}I-ubiquitin were as noted by Jahngen et al (13,14) and Haas and Bright (15). Beef lens tissue supernatants were prepared as described (16). Assays to monitor the ability of lens epithelium and cortex soluble proteins to catalyze the formation of ubiquitin conjugates were done as described (14,16).

RESULTS

Cultured rabbit lens epithelial cell mediated ATP-dependent conjugation of ubiquitin to lens proteins

Most proteolytic activities are highest in epithelial as opposed to cortex or core tissue (9,10,17,18) but many of these are even higher in cultured lens epithelial cells (19). Accordingly evidence for an intrinsic ubiquitin-lens protein conjugating capability was first sought in cultured rabbit lens epithelial cell supernatants. ^{125}I-labeled ubiquitin was added to supernatants of these cells in the presence and absence of ATP. Samples were removed at 0.5-20 min and the proteins in the resulting mixtures separated using sodium dodecyl sulfate polyacrylamide gel electrophoresis (SDS-PAGE, 20). The SDS gel stained for protein (Fig. 3, lane 1) shows patterns which are invariant over the time of the experiment. The 45 kilodalton (kDa) species in the gel is due to the carrier protein, ovalbumin. Some protein is present in the usual mass range of crystallin monomers, 17-31 kDa, but this represents only a minor fraction of the total protein present. This is as expected since it has been reported that these cells do not produce significant levels of crystallins. A very small proportion of the proteins are of high molecular mass, >200 kDa.

Ubiquitination of the cultured rabbit lens epithelial cell proteins is clearly observed in the auto radiogram (Fig. 3, lanes 2 and 3). The most prominent ^{125}I-ubiquitin conjugated proteins are at 17 and >200 kDa. Bands corresponding to ^{125}I-containing conjugates are also seen from 26 to 65 kDa. Conjugates with molecular mass below 23 kDa and within the 100-200 kDa range are also indicated, but due to marked heterogeneity these are less well resolved. All ^{125}I-containing moieties increase in intensity with time under these conditions.

In the ATP-depleted controls, only trace amounts of ^{125}I-containing proteins are observed (Fig. 3, lanes 4 and 5.). In contrast with the assays containing ATP, there is no time related increase in intensity of these bands. The most optically dense bands are at 17 and 54 kDa. These are possibly due, respectively, to diubiquitin and a small amount of ubiquitin-ovalbumin complex, formed upon storage of ^{125}I-ubiquitin with ovalbumin as carrier. The 54 kDa species is regularly observed in control assay of ^{125}I-ubiquitin and ovalbumin without any lens protein.

Conjugation of ^{125}I-ubiquitin to lens proteins using the intrinsic conjugation system and endogenous substrates in supernatants of lens tissue of different ages and developmental stages

As compared with the SDS gel of cultured lens epithelial cell supernatant shown above, protomers of crystallin 2 proteins 17-31 kDa dominate the protein profile of beef lens epithelium tissue (Fig. 4, lane 1). This is expected since crystallins are the major gene products of developing and mature lens cells.

Many ubiquitin-lens protein conjugates were formed by young and older beef lens epithelial tissue supernatants to which ^{125}I-ubiquitin was added (Fig 4., lanes 3-5). Conjugate formation was dependent upon ATP (lane 2) and accumulation increased with time (compare lanes 3 and 4). Ubiquitin conjugation to lens proteins, as judged by overall incorporation of ^{125}I-ubiquitin into species of mass greater than 8.5 kDa, was less extensive in

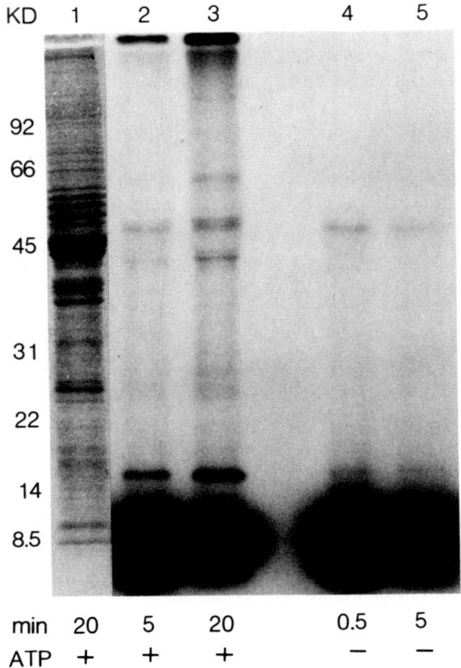

Figure 3. SDS-PAGE separation of ^{125}I-ubiquitin-protein conjugates formed by a cultured rabbit lens epithelial cell supernatant. Conjugation of ^{125}I-ubiquitin was achieved in assays containing 2 umol of Tris-HCl (pH 7.6), 250 nmol of $MgCl_2$, 50 nmol of dithiothreitol, 1 ug of ubiquitin, 500 nmol of creatine phosphate, 100 nmol of ATP, 10 ug of creatine phosphokinase, 180 ug of rabbit lens protein and 1 x 10^6 cpm of ^{125}I-ubiquitin in a final volume of 50 ul. Assays were run at 37 C and terminated with 2x Laemmli buffer (20). The SDS gel was 13% acrylamide. Lane 1 shows the Coomassie Blue stained protein profile. ^{125}I-containing conjugates were revealed by autoradiography. Lane 2 - 5 min; lane 3 - 20 min. Lanes 4 and 5 are ATP-free assays. Creatine phosphate, ATP and creatine phosphokinase were deleted and replaced with 50 nmol of 2-deoxyglucose and 4 units of hexokinase. Lane 4 - 0.5 min; lane 5 - 5 min.

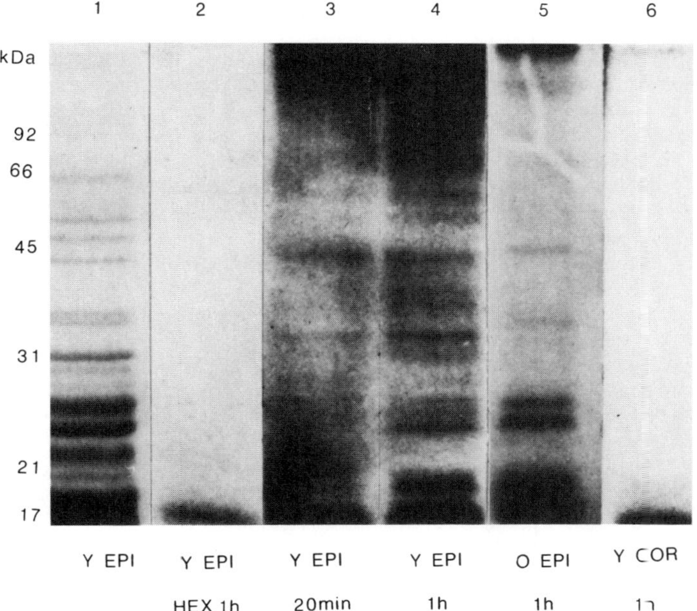

Figure 4. SDS PAGE of beef lens epithelial and cortex tissue supernatants. Assay conditions were as described for Fig. 3. Lane 1 shows the Coomassie Blue attained protein profile. Lanes 2 - 6 are autoradiograms of ^{125}I-ubiquitin-lens protein conjugation assays using young and old lens epithelium as well as young cortex as the source of substrate and conjugating enzymes. Lane 2 - hexokinase control for 1h; lane 3 - 20 min; lanes 4 - 6 - 1h. Y epi = young epithelium; O epi = old epithelium; y cor = young cortex.

older as compared with young lens epithelial tissue (lanes 5 and 4). The efficiency of ubiquitin conjugation was markedly diminished in older lens tissue (cortex, lane 6) from a young lens as contrasted with epithelium from either a young or old lens.

DISCUSSION

Damaged and altered proteins constitute a majority of the protein in more mature and aged lens tissue. In most cell systems examined damaged or altered proteins are selectively and rapidly degraded by intracellular proteases. However, aging of the lens tissue is associated with extensive protein accumulation, aggregation and eventual precipitation in eye lens cataracts (4,5). The data presented here suggest that such attenuated protein degradation may in part be due to decrements in the processes necessary to initiate proteolysis via the ubiquitin dependent pathway.

Ubiquitin-lens protein conjugates are formed in vitro incultured lens cells and lens tissue supernatants

It has been established that ubiquitination of proteins is necessary for their rapid removal by cellular proteases (for review, see 7). It appeared plausible that ubiquitin conjugation may have similar roles in lens as in the other cell types used to make this observation. High molecular weight (HMW) protein aggregates and/or HMW ubiquitin-protein conjugates are frequently the preferred substrates for proteolysis (21-24). However, during maturation and aging or upon inhibition of the ubiquitin pathway, only depressed rates of degradation of high molecular weight ubiquitin-protein conjugates are obtained in normal and transformed cells (22,23,25). The importance of the ubiquitin-dependent proteolytic pathway is suggested by the observation that high molecular weight protein aggregates are produced during aging and upon photoirradiation of the lens (4,5,26-28). Thus, impairment of this initiating event in proteolysis might be related to the observed accumulation rather than degradation of these damaged lens proteins upon aging.

Previous work from this laboratory has shown that the lens contains substantial supplies of endogenous ubiquitin (14), and that lens proteins are good substrates for ubiquitin conjugation catalyzed by the reticulocyte ubiquitin conjugating system (16). The multitude of ^{125}I-ubiquitin lens protein conjugates observed in the cell-free assays reveals that, in addition to providing substrate proteins for ubiquitin conjugating enzymes, the cultured lens cells and epithelial tissue also contain an active endogenous ubiquitin-lens protein conjugating capacity (Figs. 3 and 4). This is corroborated by immunological detection of ubiquitin-lens protein conjugates formed in vivo in lens preparations which had no exogenous additions (14).

Although Figures 3 and 4 are derived from experiments on different species, the pattern of ^{125}I-ubiquitin conjugates formed bear many similarities. The appearance of very high molecular weight conjugates (>200 kDa) in the cell-free assays of cultured epithelial lens cell supernatants (Fig. 3, lane 3) is consistent with conjugate pattersn observed when conjugation enzymes were provided by lens tissue (Fig. 4), when lens tissues and cells were immunologically probed for endogenous conjugates (14), and with conjugate patterns observed in culture IMR-90, CV-1, Friend (15), ts85 cells (29), as well as reticulocytes (15,30) and erythrocytes (15).

The exquisite selectivity of the ubiquitin-lens protein conjugation system is suggested by several observations. Extensive formation of HMW ^{125}I-ubiquitin-lens protein conjugates is observed in cultured epithelial cells (Fig. 3, lane 3) and in lens tissue supernatants (Fig. 4, lane 4). While there are some proteins present >150 kDa, these do not comprise a majority of proteins in cultured epithelial cells (Fig. 3, lane 1). Indeed, in no case does the predominant conjugate correspond to a major protein of similar molecular mass.

The data in Fig. 4 suggest that ubiquitin conjugation is catalyzed least effectively in older (cortex) tissue supernatants. This is consistent with studies which show that in older or more mature lens tissue, proteolytic capacity and a variety of enzyme activities which aid in antioxidant function and glucose metabolism are in a state of attenuated activity (2,8,10,17,31). A decrease in the amount of ubiquitin conjugates formed with epithelial tissue from older animals indicates that the ubiquitin conjugating system may suffer an age-related decrease in efficiency.

Biological importance of proteolysis during lens aging

The crystallins, the predominant lens proteins, sustain substantial modification throughout life. Similar proteolytic machinery exists in young, and to some extent old, lens cells (1,2,8-10,17,32-35) as is available in many other cell types. Thus, the accumulation of postsynthetically altered crystallins in maturing aged lens tissue remain to be explained.

Accessible alpha amino termini of putative protein substrates seem to be required for both initiation of ubiquitin/ATP-dependent proteolysis (24) and for aminopeptidase action (36). Over 80% of cellular proteins have unavailable alpha amino termini (24), including the alpha, beta and gamma crystallins. Furthermore, if endopeptidase activity begins only subsequent to marking of the substrate protein by ubiquitination, then endopeptidase activities will remain ineffective against crystallins. especially those crystallins which have inaccessible alpha-amino termini. This could explain the stability of the crystallins. especially in younger tissues.

With age or upon photolysis amino termini become available. Along with denaturation this should provide a large and diverse substrate population for the ubiquitin proteolytic pathway and for other lens proteases. Why are these altered molecules not degraded? The data in Jahngen et al. (14,16) and that shown in Figures 3 and 4 suggest that, whereas substrates in older tissue may be available, the ubiquitinating capability and some of the ATP-independent endopeptidase activities of the older tissues is markedly compromised (2,8-10). Together with markedly reduced aminopeptidase activity (8,17), this could result in attenuated proteolysis in older as compared with younger lens tissue (10). Thus, diminished proteolytic capabilities as well as enhanced covalent crosslinking of damaged protein substrates might preclude sufficient proteolysis and the timely removal of damaged proteins in old lens tissue. This idea is consistent with observations that oxidatively stressed, aged and protease-inhibited red blood cells accumulate protein aggregates such as Heinz bodies (23,37,38). Similar reasoning may explain the accumulation of ubiquitin conjugates in a variety of age-related neurological diseases (39,40). Since the proteases are also proteins, it is not unexpected that they sustain similar age-related changes as occur to the crystallins. This has recently been confirmed in experiments in which lens aging was simulated by UV irradiation (26-28). It is of interest that such UV-induced age-related insult to lens proteins may be attenuated by elevated dietary ascorbate (27).

ACKNOWLEDGEMENTS

We acknowledge Drs. Finley, Varshavsky and Ciechanover for discussions and contribution of ubiquitin and Ester Epstein and Libby Jensen for word processing. This work was supported by grants from USDA Contract No. 53-3K06-5-10, NIH, Fight for Sight, the Daniel and Florence Guggenheim Foundation and the Massachusetts Lions Eye Research Fund, Inc.

REFERENCES

1. A.M.J. Blow, 1979, Proteolysis in the lens, In: "Proteninases in Mammalian Cells and Tissues," A.J. Barrett, ed., North-Holland Publishing Co., Amsterdam.

2. A. Taylor and K.J.A. Davies, 1987, Protein oxidation and loss of protease activity may lead to cataract formation in the aged lens, Free Radical Biol. Med. 3:371.

3. J. Papacoustaninou, 1967, Molecular aspects of lens cell differentiation, Science 156:338.

4. J.A. Thomson. T.P. Hum and R.G. Augusteyn, 1985, Reconstructing normal alpha-crystallin from the modified cataractous protein, Aust. J. Exp. Biol. Med. Sci. 63:563.

5. J.J. Harding, 1981, Changes in lens proteins in cataract. In: "Molecular and Cellular Biology of the Eye lens," H. Bloemenda, ed., J. Wiley and Sons, New York.

6. A. Hershko, 1983, Ubiquitin: roles in protein modification and breakdown, Cell 34:11.

7. M. Rechsteiner, 1987, Ubiquitin-mediated pathways for intracellular proteolysis, Ann. Rev. Cell Biol. 3:1.

8. A. Taylor, M. Daims, J. Lee and T. Surgenor, 1982, Identification and quantification of leucine aminopeptidase in aged normal and cataractous human lenses and ability of bovine lens LAP to cleave bovine crystallins, Curr. Eye Res. 2:47.

9. H. Yoshida, T. Murachi and I. Tsukahara, 1984, Limited proteolysis of bovine lens alpha crystallin by calpain, a Ca^{2+} -dependent cysteine proteinase isolated from the same tissue, Biochem. Biophys. Acta 798:252.

10. K.R. Fleshman and B.J. Wagner, 1984, Changes during aging in rat lens endopeptidase activity, Exp. Eye Res. 39:543.

11. R. van Heyningen, 1978, Lens neutral proteinase, Interdiscpl. Topics Geront. 12:232.

12. J.H. Jahngen and A. Taylor, 1985, Aging decreases eye lens protein-ubiquitin conjugation, Fed. Proc. 44:1226.

13. A. Taylor, A.L. Haas, J.H. Jahngen and J. Blondin, 1986, Age related changes of in vivo eye lens ubiquitin and ubiquitin conjugates and plausible role in cataract formation, Invest. Ophthalmol. Visual Sci. 27:277.

14. J. H. Jahngen, A.L. Haas, A. Ciechanover, J. Blondin, D.E. Eisenhauer and A. Taylor, 1986, The eye has an active ATP-dependent ubiquitin protein conjugation system, J. Biol. Chem. 261:13760.

15. A.L. Haas and P.M. Bright, 1985, The immunochemical detection and quantification of intracellular ubiquitin-protein conjugates, J. Biol. Chem. 260:12464.

16. J.H. Jahngen, D.E. Disenhauer and A. Taylor, 1986, Lens proteins are substrates for the reticulocyte ubiquitin conjugation system, Curr. Eye Res. 5:725.

17. A. Taylor, M.J. Brown, M.A. Daims and J. Cohen, 1983, Localization of leucine aminopeptidase in normal hog lens by immunofluorescence and activity assays, Invest. Ophthalmol Visual Sci. 24:1172.

18. K.R. Fleshman, J.W. Margolis, S.-C. J. Fu and B.J. Wagner, 1985, Age changes in bovine lens endopeptidase activity, Mech. Aging Develop. 31:37.

19. D.A. Eisenhauer, J.J. Berger, C.Z. Peltier and A. Taylor, 1988, Protease activities in cultured beef lens epithelial cells peak and then decline upon progressive passage, Exp. Eye Res. 46:579.

20. U.K. Laimmli, (1970) Cleavage of structural proteins during the assembly of the head of the bacteriophage T4, Nature 227:680.

21. A. Hershko, E. Leshinskey, D. Ganoth, and H. Heller, 1984, ATP-dependent degradation of ubiquitin-protein conjugates, Proc. Nat'l. Acad. Sci. USA 81:1619.

22. R. Hough and M. Rechsteiner, 1986, Ubiquitin-lysozyme conjugates, J. Biol. Chem. 261:2391.

23. R. S. Daniels, V.C., Worthington, E.M. Atkinson and A.R. Hipkiss, 1980, Proteolysis of puromycin-peptides in rabbit reticulocytes: detection of a high molecular weight oligopeptide proteolytic substrate, FEBS Lett. 113:245.

24. A. Hershko, H. Heller, E. Eytan, G. Kaklij and I.A. Rose, 1984, Role of the alpha-amino group of protein in ubiquitin-mediated protein breakdown, Proc. Nat'l. Acad. Sci. USA 81:7021.

25. S. Speiser and J.D. Etlinger, 1982, Loss of ATP-dependent proteolysis with maturation of reticulocytes and erythrocytes, J. Biol. Chem. 257:14122.

26. J. Blondin and A. Taylor, 1987, Measures of leucine aminopeptidase can be used to anticipate uv-induced age-related damage to lens proteins: ascorbate can delay this damage, Mech. of Aging Develop. 41:39.

27. J. Blondin, V. Baragi, E. Schwartz, J.A. Sadowski and A. Taylor, 1986, Delay of uv-induced eye lens protein damage in guinea pigs by dietary ascorbate, J. Free Radicals Biol. Med. 2:275.

28. A. Taylor, J.H. Jahngen, J. Blondin and E.G.E. Jahngen, Jr., 1987, Ascorbate delays ultraviolet-induced age-related damage to lens protease and the effect of maturation and aging on the function of the ubiquitin-lens protein conjugating apparatus. In: "Proteases in Biological Control and Biotechnology," G. Long and D. Cunningham, eds., Alan R. Liss, Inc., New York.

29. A. Ciechanover, D. Finley and A. Varshavsky, 1984, Ubiquitin dependence of selective protein degradation demonstrated in the mammalian cell cycle mutant ts 85, Cell 37:57.

30. A. Hershko, E. Eytan, A. Ciechanover and A.L. Haas, 1982, Immunochemical analysis of the turnover of ubiquitin-protein conjugates in intact cells, J. Biol. Chem. 257:13964.

31. A.Z. Reznick, A. Covrat, L. Rosenfelder, S. Shpund and D. Gershon, 1985, Defective enzyme molecules in cells of aging animals are partially denatured, totally inactive, normal degradation intermediates. In: "Modification of Proteins During Aging," R.C. Adelman and E.E. Dekker, eds., Alan R. Liss, Inc., New York.

32. B.J. Wagner, S.-C. J. Fu, J.W. Margolis, and K.R. Fleshman, 1984, A systhetic endopeptidase substance hydrolyzed by the bovine lens neutral proteinase preparation, Exp. Eye Res 38:477.

33. O.P. Srivastava and B.J. Ortwerth, 1983, Isolation and characterization of a 25K serine proteinase from bovine lens cortex, Exp. Eye Res. 37:597.

34. R.W. Gracy, K.U. Yuksel, M.L. Chapman, J.D. Cini, M. Jahani, H.S. Lu, B. Oray and J.M. Talent, 1985, Impaired protein degradation may account for the accumulation of "abnormal" proteins in aging cells. In: "Modification of Proteins During Aging", R.D. Adelman and E.E. Dekker, eds., Alan R. Liss, Inc., New York.

35. K. Ray and H. Harris, 1985, Purification of neutral lens endopeptidase: close similarity to a neutral proteinase in pituitary, Proc. Nat'l. Acad. Sci. USA 82:7545.

36. E.L. Smith and R.L. Hill, 1960, Leucine aminopeptidase. In: "The Enzymes", vol 4, P.D. Boyer, ed., Academic Press, New York.

37. M.J. McKay, R.S. Daniels and A.R. Hipkiss, 1980, Breakdown of aberrant protein in rabbit reticulocytes decreases with cell age, Biochem J. 188:279.

38. C.C. Winterbourn, 1979, Protection by ascorbate against acetylphenyl hydrazine-induced Heinz body formation in glucose-6-phosphate dehydrogenase deficient erythrocytes, Brit. J. Haematology 41:245.

39. V. Manetto, G. Perry, M. Tabaton, P. Mulvihill, V.A. Fried, H.T. Smith, P. Gambetti and L. Autilio-Gambetti, 1988, Ubiquitin is associated with abnormal cytoplasmic filaments characteristic of neurodegenerative diseases, Proc. Nat'l. Acad. Sci. USA 85:4501.

40. S. Love, T. Saitoh, S. Quijada, G.M. Cole and R.D. Terry, 1988, Alz-50, ubiquitin and Tau immunoreactivity of neurofibrillary tangles, Pick bodies and Lewy bodies, J. Neuropath. Exp. Neurology, 47:393.

UPDATE 1990

Recent work has extended the observations mentioned above in rabbit lens epithelial cells and calf lens tissues to include culture bovine lens epithelial (BLE) cells. We have shown that histone H2A, alpha crystallin, and actin are conjugated to ubiquitin, resulting in higher molecular mass species, which are detected by anti-ubiquitin antibodies (41). These proteins are also degraded in cell-free assays containing BLE cell supernatants under physiological conditions in an ATP/Mg^{2+}-dependent manner. Additionally we show the ATP-dependent formation of HMW aggregates as well as the formation of proteolytic fragments of ^{125}I-alpha crystallin in the presence of BLE cells. These data are all consistent with ubiquitin-dependent degradation of these proteins.

Visualization of ^{125}I-alpha crystallin degradation on autoradiograms

As mentioned above, a feature of the ubiquitin-dependent proteolytic system is that substrates are conjugated with multiple ubiquitins prior to degradation and that this conjugation is ATP/Mg^{2+}-dependent. The HMW moieties of ^{125}I-alpha crystallin that accumulate with time as shown in Figure 5 (hollow arrow, lanes 2-4) may be evidence of this process. In order to establish that ATP was required for the formation of HMW aggregates as well as degradation, endogenous ATP contained in the BLE cells was removed with hexokinase. There was an ATP/Mg^{2+} dependence in the production of the HMW moieties (compare lanes 3 and 6). No HMW species were observed in the absence of BLE cell supernatant (compare lanes 3 and 9).

The generation of low molecular mass fragments with BLE cell supernatants can also be seen in Figure 5. The autoradiogram obtained after SDS polyacrylamide gel electrophoresis shows a time-dependent generation of proteolytic fragments (PF, Fig. 5, lanes 2-4) that is also ATP/Mg^{2+}-dependent.

Some preparations of hexokinase are known to contain proteolytic capabilities (A.T. and J.J.H., unpublished data). With hexokinase and the BLE cell supernatant, there was indication of proteolysis (PF, Fig. 5, lanes 5-7) above the 0 hr value. In another control when hexokinase was used alone, there was evidence of the time-dependent formation of low molecular mass species less than alpha crystallin but greater than those fragments that run at the dye front on the gel (short arrow, Fig. 5, lanes 8-10). This indicates that the hexokinase preparation used was contaminated with a (some) protease(s). As compared with lanes 5-7 in Figure 5, lanes 8-10 give the impression of higher levels of proteolysis (i.e., more radiolabelled polypeptides are observed below 10 kDa). This could occur because proteins present in the BLE cell preparation spare the radiolabelled crystallin substrate and/or because peptidases present in the BLE cell preparation digest the polypeptides formed to amino acids. Some proteolytic fragments are observed in Figure 5, lanes 5-7 (also see ref. 19).

<u>Quantification of degradation of histone H2A, alpha crystallin, and actin by measurement of release of TCA-soluble fragments</u>

Proteolysis by BLE cell supernatant using various iodinated substrate proteins is shown in Table I. The proteins were degraded with differing efficiency by BLE cell and reticulocyte supernatants. For example with BLE cell supernatant in the presence of ATP/Mg^{2+}, 25% of the histone H2A was degraded whereas in the absence of ATP, only 9.3% of the substrate was hydrolyzed. In contrast, the reticulocyte supernatant hydrolyzed 16% of the H2A in the presence of ATP/Mg^{2+}. In the absence of ATP/Mg^{2+}, approximately 3.1% of the N2A was degraded. With alpha crystallin, the average level of proteolysis with the BLE cell supernatant was 2.3% (with ATP/Mg^{2+}. In the absence of ATP, there was only 4.2% of the alpha crystallin degraded. Using BLE cell and reticulocyte preparations, actin was degraded only in the presence of ATP/Mg^{2+}. The exclusive proteolysis of actin with ATP/Mg^{2+} and the enhanced degradation of H2A and alpha-crystallin in the presence of ATP/Mg^{2+} indicates the requirement of ATPMg^{2+} and putatively ubiquitin for action of the BLE cell and reticulocyte supernatant cytosolic proteolytic pathway. It is interesting to note that the net enhancement supernatant is 11.5, although the absolute amount of proteolysis is less than is observed with reticulocyte supernatant.

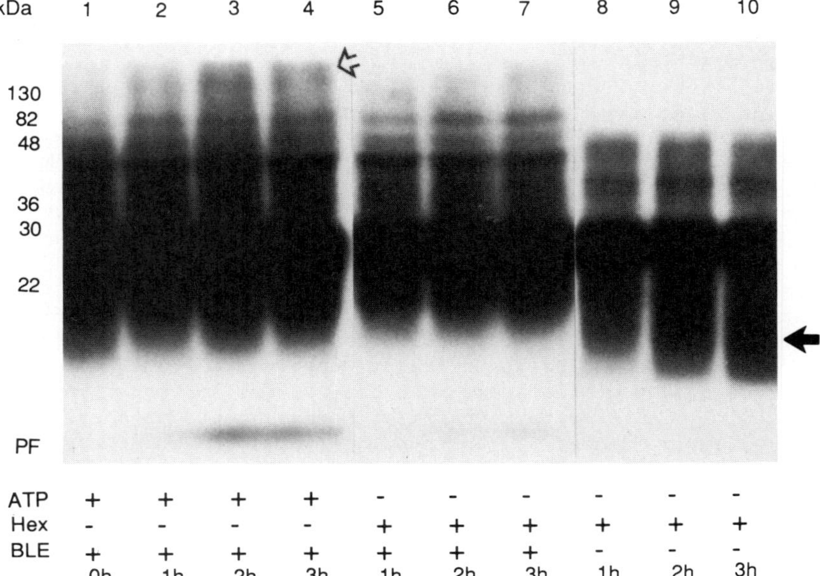

Figure 5. Autoradiogram of separation of alpha-crystallin following hydrolysis with BLE cell supernatant. In a final volume of 50 ul, 40 mM Tris-HCl (pH 7.6), 5mM $MgCl_2$, 1 mM DTT, 10 mM creatine phosphate, 2 mM ATP, 5 ug og creatine phosphokinase, and 2-5 x 10^4 cpm of the various ^{125}I-labelled protein substrates were incubated with 600 ug BLE cell supernatant for up to 3 h at 37 C. In order to observe the ^{125}I fragments produced during proteolysis, aliquots of the assays (1-4 x 10^4 cpm) were denatured by boiling for 2 min in sample buffer and layered on top of 12.5% SDS polyacrylamide gels (20) with a 4% stacker gel. Gels were dried without staining and autoradiograms obtained following exposure at -80 C with Kodak XAR-5 film using a DuPont Lightning Plus intensifying screen. Approximate molecular masses of prestained standards from Bio Rad were phosphorylase B, 130 kDa, BSA, 82 kDa; ovalbumin, 48 kDa; carbonic anhydrase, 36 kDa; soybean trypsin inhibitor, 30 kDa; lysozyme, 22 kDa. PF=proteolytic fragments; short arrow = low molecular mass proteins; hollow arrow = high molecular mass proteins.

TABLE I

DEGRADATION OF ^{125}I-LABELLED PROTEINS BY BOVINE LENS EPITHELIAL CELL AND RETICULOCYTE SUPERNATANTS

	BLE cell supernatant			Retic supernatant		
	+ATP	-ATP	+ATP/-ATP	+ATP	-ATP	+ATP/-ATP
Actin	2.9	0	TD	8.0	0	TD
Histone H2A	25.2	9.3	2.7	16.0	3.1	5.2
Alpha-crystallin	2.3	0.2	11.5	10.8	4.2	2.6

Assay conditions: Liberation of TCA-soluble fragments was determined with 600 ug BLE cell supernatant or 20 ul of reticulocyte supernatant for 3 h. The reaction was terminated by the addition of 400 ul of 10% (w.v) bovine serum albumin, followed by 100 ul of 100% (w.v) TCA. After allowing precipitation at 4 C for 30 min, the protein was centrifuged at 12,000 x g for 15 min and the total cpm in each reaction tube determined by gamma counting. Aliquots of the supernatant were taken to determine the TCA-soluble radioactivity, and the pellet drained and counted to ascertain the TCA-insoluble radioactivity. To determine the extent of ATP-independent proteolysis in BLE cell supernatants, the creatine phosphate, ATP, and reatine phosphokinase were replaced with hexokinase (4 units) and 2-deoxyglucose (10 mM). Results are expressed as per cent degradation as determined in the TCA supernatant (sup). The following calculation was used:

$$\frac{\text{exptl TCA sup cpm at 3 h - blank TCA sup cpm at 3 h}}{\text{total cpm in assay - blank TCA sup cpm at 3 h}}$$

Blank for +ATP consisted of ATP, creatine phosphokinase and creatine phosphate; blank for -ATP contained hexokinase and 2-deoxyglucose. All values are the average of triplicates for each experiment, and each experiment was repeated at least twice. TD = totally dependent upon +ATP.

DISCUSSION

Of the three proteins tested for proteolysis, the nucleosomal protein histone H2A was degraded most effectively by the BLE cell supernatant (25% in the presence of ATP versus 9.3% in the absence of ATP). This is one of the most rapid proteolytic processes observed to date in cell-free assays and may be considered important in light of lens development. The lens epithelial cell matures into an elongated fiber cell. As the lens fiber cell undergoes terminal differentiation, it loses its nuclei as well as other organelles (42). The nucleosomal H2A and associated DNA are lost from the nucleus during pycnosis which follows a strict temporal of the nuclear membrane and chromatin, and the destruction of the fine structure of the nuclei (43). If this pattern is perturbed, resulting in an incomplete denucleation, cataractous changes may occur in the lens (44). It would appear that ATP/Mg^{2+} and presumably ugiquitin-dependent degradation of histone H2A are required during this denucleation process.

Degradation of alpha crystallin to TCA insoluble proteolytic fragments was also observed under physiological conditions with supernatants from cultured BLE cells in the presence of ATP. Quantification of low molecular mass fragments by scanning of the autoradiogram shown in Figure 5 (PF) or by counting the TCA-soluble radioactivity (Table I) showed that approximately 2.0% of the total radioactivity was degraded in an ATP-dependent manner in 3 h. Alpha crystallin has been shown to be a substrate for the lens proteasome (35,45,46). Lens proteasome is similar to a major proteolytic activity in many eucaryotic cells (47-49) which has a ubiquitin-dependent component (50,51). It is probable that the proteasome in BLE cells is involved in degradation of alpha crystallin.

It has previously been shown that coincident with proteolysis is an accumulation of HMW ugiquitin-protein conjugates in partially purified reticulocyte preparations (22,52,53). The observation of the time-dependent accumulation of HMW species in the presence of ATP (Fig. 5, lanes 1-4) is also consistent with ubiquitin-dependent proteolysis of alpha crystallin. The numerous HMW moieties may be alpha crystallin with varying numbers of ubiquitin molecules attached. Further implication of involvement of the ubiquitin system is given by the observation that a major ubiquitin-alpha crystallin conjugate at approximately 30 kDa is identified by anti-ubiquitin antibodies (41). Blots probed with rabbit anti-alpha crystallin antibody also react with the 30 kDa protein confirming that the 30 kDa protein is a ubiquitin-alpha crystallin conjugate (data not shown).

Actin was degraded only in the presence of ATP/Mg^{2+}. The observation of an actin conjugate plausibly containing multiple ubiquitin molecules (41), as detected by anti-ubiquitin antibodies, suggests that ubiquitin is involved in the proteolysis of actin. Actin has also been shown to be a substrate for protease(s) in reticulocyte supernatant (Table I, ref. 54) and was found conjugated to ubiquitin in insect flight muscle (55). Actin filaments frequently form aggregates (stress fibers) that may serve a function in lens cell movement and elongation (56). This important role of actin in BLE cells and lens tissue may teleologically justify its actin degradation that occurs in BLE cells seems to occur via the ubiquitin-dependent pathway.

CONCLUSION

We have demonstrated that the lens has a fully functional ubiquitin conjugating capability which is the first step in ubiquitin-dependent proteolytic system. A hallmark of degradation via the ubiquitin-dependent pathway is a requirement for ATP/Mg^{2+}. Endogenous lens proteins, alpha-crystallin, actin, and H2A, are degraded in an ATP/mG^{2+}-dependent fashion by BLE cell proteases. Taken together these data suggest that much of the degradation of these proteins is accomplished via the ubiquitin-dependent proteolytic pathway. Aging and/or maturation results in compromises of aspects of this proteolytic process. Some ubiquitin conjugates may not form as effectively, and others may be constructed inappropriately. Combined with decreased proteolytic capabilities, these phenomenon may be responsible in part for the accumulation of damaged proteins in cataract.

REFERENCES

41. J.Jahngen-Hodge, E. Laxman, A. Zuliani, and A. Taylor, Evidence for ATP and ubiquitin-dependent degradation of proteins in cultured bovine lens epithelial cells, Exp. Eye Res., in press.

42. J. Piatigorsky, 1981, Lens differentiation in vertebrates, Differentiation 19:134.

43. S.P. Modak and S.W. Perdue, 1970, Terminal lens cell differentiation, Exp. Cell Res. 59:43.

44. T. Kuwabara and M. Imaizumi, 1974, Denucleation process of the lens, Invest. Ophthalmol. 13:973.

45. K. Murakami, J.H. Jahngen, S. Lin, K.J.A. Davies, and A. Taylor, 1990, Lens proteasome shows enhanced rates of degradation of hydroxyl radical modified alpha-crystallin, Free Radicals Biol. Med. 8:217.

46. B.J. Wanger, J.W. Margolis, P.N. Farnsworth, and S.-C. Fu, 1982, Calf lens neutral proteinase activity using calf lens crystallin substrates, Exp. Eye Res. 35:293.

47. B. Dahlman, L. Keuhn, M. Rutschmann, and H. Reinauer, 1985, Purification and characterization of a multicatalytic high-molecular-mass proteinase from rat skeletal muscle, Biochem. J. 228:161.

48. M.J. McGuire and G.N. DeMartino, 1986, Purification and characterization of a high molecular weight proteinase (macropain) from human erythrocytes, Arch. Biochem. Biophys. 262:273.

49. A.P. Arrigo, K. Tanaka, A.L. Goldberg, and W.J. Welch, 1988, Identity of the 19S "prosome" particle with the large multifunctional protease complex of mammalial cells (the proteasome), Nature 331:192.

50. J. Driscoll and A.L. Goldberg, 1990, The proteasome (multicatalytic protease) is a component of the 1500-kDa proteolytic complex which degrades ubiquitin-conjugated proteins, J. Biol. Chem. 265:4789.

51. R. Hough, G. Pratt, and M. Rechsteiner, 1987, Purification of two high molecular weight proteases from rabbit reticulocyte lysate, J. Biol. Chem. 262:8303.

52. A. Hershko and H. Heller, 1985, Occurrence of a polyubiquitin structure is ubiquitin-protein conjugates, Biochem. Biophys. Res. Commun. 128:1079.

53. M. Magnani, V. Stocchi, L. Chiarantimi, G. Serafini, M. Dacha, and G. Foranini, 1986, Rabbit red blood cell hexokinase, J. Biol. Chem. 261:8327.

54. A. Mayer, N.R. Siegel, A.L. Schwartz, and A. Ciechanover, 1989, Degradation of proteins with acetylated amino termini by the ubiquitin system, Science 244:1480.

55. E. Ball, C.C. Karlik, C.J. Beall, D.L. Saville, J.C. Sparrow, B. Bullard, and E.A. Fryberg, Arthrin, 1987, a myofibrillar protein of insect flight muscle, is an actin-ubiquitin conjugate, Cell 51:221.

56. G.C.S. Ramaekers, M.W.A.C. Hukkelhoven, A. Groeneveld, and H. Bloemendal, 1984, Changing protein patterns during lens cell aging in vitro, Biochem. Biophys. Acta 700:221.

DEVELOPMENT AND AGE-ASSOCIATED CHANGES IN EXTRACELLULAR MATRIX OF THE HUMAN EYE

Amos Ben-Zvi, Merlyn M. Rodriques, Sankaran Rajagopalan, and Leslie S. Fujikawa

Laboratory of Eye Pathology and Cell Pathology
University of Maryland at Baltimore, School of Medicine
Baltimore, Maryland 21201

Age-associated changes in the embryo occur from the moment of conception as a part of development. The cornea is no exception in reflecting the developmental changes in the process of growth and the wear and tear of the aging process. It is often possible to estimate the age of the cornea by histological study including electron microscopy, or by the newly developed immunohistochemical methods of monoclonal antibodies against developmentally regulated corneal antigens, using the immunofluorescent or immunoperoxidase staining. The morphologic age-related manifestations in the eye are well known, but details of the use of monoclonal antibodies as a tool detecting specific antigens of the cornea including the corneal extracellular matrix will be presented here.

Aging: The cornea manifests a considerable amount of age associated changes (Kuwabara, 1976; Hogan et al, 1962; Yanoff et al, 1975). Since most of these changes occur in the peripheral cornea, the transparency of the central cornea is preserved.

Epithelium and epithelial basement membrane: The thickness of the basal lamina beneath the basal epithelium increases with age, and the young flat basal surface becomes irregularly wavy. Sometimes the basal epithelium in the limbus area extends into Bowman's membrane; often the cytoplasm of the epithelium cells contain iron particles that can form brown inclusion bodies in the horizontal palpebral line: Hudson-Stahli line in elderly individuals.

Bowman's membrane: Unlike the basal lamina beneath the basal epithelium, the thickness and the general appearance of Bowman's membrane remains unchanged in the aging process but "arcus senilis" deposits of fatty substances are formed in the peripheral zone of Bowman's membrane. Often small foci of calcification are found in Bowman's membrane as well as in the thickened basal lamina of the epithelium. Moreover, in some aged eyes small basophilic bodies, believed to be collagen metabolites are deposited peripheral to the termination of Bowman's membrane.

Descemet membrane: Descemet's membrane thickness increases with age. The membrane continues to thicken from 1.0 micron at the time of birth to 15 microns in old age (Cogan et al, 1971). After infancy the anterior part of Descemet's membrane becomes coarse and the distinction of two layers becomes prominent with age. The basal lamina substance in the anterior layer form bundles of short filaments. In the cornea

The Effects of Aging and Environment on Vision
Edited by D.A. Armstrong *et al.*, Plenum Press, New York, 1991

with "arcus senilis," fatty changes and deposits of cholesterol crystals may occur in this layer. In aged peripheral Descemet's membrane numerous warts, Hassal-Henle bodies, consisting of basal lamina, short collagen fibers and wide banded fibers are found.

Endothelium: The embryonal cells are cuboidal. They are relatively flat at birth and show further flattening with aging (Cogan et al, 1971, 1976). The endothelium, unlike the epithelium, shows no mitotic activity in the adult cornea and the number of endothelial cells is reduced by 30% in elderly subjects. There is a senile enlargement of cells (Laing et al, 1975; Bourne et al, 1984) that might play a major role in covering denuded areas in the wound healing process. The endothelial cells contain filaments, micro-organelles and in the peripheral zone the cells contain melanin pigment granules. Central and vertical pigmented endothelial cells may form Krukenberg's spindle.

Stroma: The stroma of the cornea remains unchanged in elderly people, however, ultrastructural studies reveal that filaments, fragments of cell debris and elastic fibers are increased (Kanai et al, 1971). Nerve fibers are more frequently found with age but the size of the collagen fibers is not altered with age. Accumulation of excessive basal laminae substance occurs around the keratocytes, mainly in the posterior half and an abundant population of resting wandering cells (Kuwabara, 1976) that can become phagocytic cells.

Arcus senilis: Opacification of the peripheral zone, forming a discrete ring sparing a clear zone at the limbus, is defined as "arcus senilis" or arcus lipidosis (Cogan et al, 1959). It contains a mixture of cholesterol esters, neutral lipids and phospholipids (Andrew et al, 1961; Tschetter, 1966) deposited extracellularly between collagen lamellae. The lipids may have a vascular origin and both Bowman's and Descemet's membranes are involved in these age related fatty changes. Arcus senilis-like changes are observed in hypercholesterolemic patients and animals (Cogan et al, 1959) and histological similarities to atheroma of the artery have been observed.

Developmental age-associated changes in the human cornea
Developmental manifestations: Embryology: General Consideration.

After the lens vesicle separates from the surface ectoderm at 6 week's gestation, the mesoderm surrounding the optic cup sends a wave between the surface ectoderm and lens vesicle to form a layer of mesothelial cells, the corneal endothelium. This is rapidly followed by a second wave between the surface ectoderm and endothelium which forms the corneal stroma (Kiel et al, 1939). The lens is necessary for the formation of the endothelium and Descemet's membrane, since the removal of the lens vesicle results in the formation of sclera-like tissue instead of corneal tissue, with absence of endothelium and Descemet's membrane as shown by Zinn (1970). At 7 weeks gestation the endothelium is a loosely arranged layer, one to two cells thick, with the formation of a single layer of cells a week later. At 3 months, intercellular junctions appear and active transport has begun. At 4 months' gestation, a filamentous irregular basement membrane develops along the basal site of the endothelial cells. The addition of lamellae results in a more homogeneous layered membrane with about 100 nm banding by the 8th month. At birth the basement membrane consists entirely of vertically banded material (Smelser et al, 1972; Wulle, 1972) and is about the same thickness as an endothelial cell. This banded portion is the embryonic Descemet's membrane and the granular posterior zone is secreted by the endothelium throughout life (Chi et al, 1958; Cogan et al, 1971; Hogan et al, 1961).

Ultrastructural components

Epithelium: The cuboidal epithelium cells are loosely attached to each other by gap junctions and rest on the basal laminae, which is present at the earliest stage. The cytoplasm consists of an electron lucent matrix containing a few mitochondria and rough endoplasmic reticulum. Fine filaments and glycogen granules begin to accumulate in the cytoplasm (human on the 50th day). The superficial cells are loosely attached to the basal cells by sparse desmosome.

Bowman's membrane: Fine filaments measuring about 10-20 nm in diameter are formed beneath the subepithelial basal laminae before the stroma cell invades into the area. The filaments increase in number and length and this layer becomes Bowman's layer. The early collagen formation in the developing chick cornea has been well documented (Hay et al, 1969) and the layer becomes distinct when the cellular stroma is formed. The thick (1.0 nm) acellular Bowman's membrane is present in the eye of the human newborn infant. It increases in thickness to 1.2 nm, the adult size soon after.

Endothelium: Cells which are forming a cluster at the limbal zone begin to migrate centrally along the ectodermal layer. The cells are mainly polygonal in shape and they rapidly proliferate with mitosis, forming the stroma. Cells at the posterior margin of the rudimentary stroma become spindle shaped and separate the cornea from the anterior chamber which contains vascular connective tissue. These cells gradually develop junctions and become the endothelium, and the development of the junctions have been described in detail (Wulle et al, 1974). A basal lamina layer, which has a laminated filamentous appearance, forms beneath the developing endothelium and increases in thickness. This thick basement membrane is Descemet's membrane. Completion of the differentiation of the endothelium occurs around the 50th day in the human embryo. Descemet's membrane in the human at birth is about 0.5-0.7 pm thick. The endothelium is about 10 pm in height.

Stroma: The stromal tissue rapidly increases in thickness with the increase of stroma cells and the formation of collagen fibers. Developing stromal cells have numerous processes with form junctions with neighboring cells, and contain extremely rich endoplasmic reticula and Golgi apparatus. Lumina of the cisternae are enlarged and contain a fine granular substance. Short collagen fibers which are uniformly small. about 30 nm in diameter, begin to accumulate rapidly in the stroma following the proliferation of the stroma cells. Fine filaments are abundantly present among collagen fibers. The collagen fibers form bundles of various sizes and are eventually arranges into lamellae especially in the posterior half. The mucopolysaccharide appears to accumulate at a later stage of the stromal development. The corneal stroma is slightly opaque at birth. Nerve fibers are frequently present among the developing stroma cells at the early developmental stage, and fine nerve fibers enter between the basal epithelium soon after. The developmental studies of the cornea (Wulle et al, 1974, Pei et al, 1971; Wulle 1974) confirmed that the primitive cornea is formed by the epithelium. The stromal connective tissue is then formed by the cells of the nonvascular tissue formed at the limbal zone. The origin of these cells has been attributed to the neural crest cell as shown by Hay (1981). Important studies on collagen formation (Johnson et al, 1974; Kanai et al, 1971; Kiele et al, 1939; Kleinman et al 1981) in the cornea have demonstrated that the corneal epithelium produces collagen fibers and the Golgi apparatus in the epithelial cell is involved in this function. Moreover, corneal collagen requires the presence of lenticular tissue in the eye as demonstrated by Zinn (1970).

Developmental and age-associated changes in human cornea
Developmental and age-associated changes in the extracellular matrix

Collagen: General consideration: Numerous criteria define a collagen and distinguish it from other proteins (Bornstein et al, 1980; Eyre, 1980). Collagen contains a triple helical segment composed of three chains that have a characteristic composition including 33% glycine, 10% proline, 10% hydroxyproline, and a variable amount of hydroxylysine. Collagen, but not other proteins, are degraded by bacterial collagenoses, which recognize and cleave the sequence gly-x-pro-gly.These as well as other features such as their x-ray pattern establish certain structural proteins as collagen and other proteins as containing collagenous domains. At present two nonstructural proteins having collagenous segments have been identified: clq a component of the complement cascade and acetylcholinesterase (Kleinman et al, 1981). Five isotopes of collagen molecules have been characterized and others probably exist (Bornstein et al, 1980). Although the interstitial collagen type I, II, and III are the products of separate genes and have unique amino acid sequences, in other respects they are quite similar. Each molecule contain a triple helical domain, roughly 300x 1.5 nm composed of three chains each 94500 daltons.

Type I collagen found in the skin, bone and tendons has two alpha-1 (I) chains and one alpha-2 (I) chain. In addition, collagen molecules containing only alpha-1 (I) chains with the composition [alpha-1 (I)]3 have been identified and are usually referred to as type I trimer. Type I trimer is found in tissues (Kleinman et al, 1981; Hay, 1981) as product of certain cultured cells (Mayne et al, 1980; Mayne et al, 1975;) and in tumors (Moro et al, 1977). This collagen is synthesized by fibroblasts, osteoblast, smooth muscle cells and epithelium (Kleinman et al, 1981).

Type II collagen is found mainly in cartilage and contains three alpha-1 (II) chains (Miller et al, 1974). This collagen is synthesized by chondrocyte, notochord cells, and neural retinal cells.

Type III collagen once thought to be the "fetal" collagen is composed of three alpha-1 (III) chains and is most prominent in blood vessels, skin and the parenchyma of internal organs (Gay et al, 1978). Many investigators consider it to be the reticulin described by histochemical methods. This collagen is synthesized by fibroblasts and myoblast.

Type IV collagen is found in basement membranes (Bentz et al, 1978; Bornstein et al 1980) and contains two distinct chains that are larger than the chains of other collagens and it has more carbohydrate, 3-hydroxyproline and hydroxylysine than other collagens.

Type IV collagen is synthesized by endothelial and epithelial cells.

Type V collagen is present in the skin, smooth muscle, placenta and bone (Kleinman et al, 1981) contains two types of chains designated A and B or alpha-2(V). The two chains occur in the same molecule in type V collagen from placenta and molecules containing only B chains have been isolated from cartilage (Kleinman et al, 1981).

Fibronectin: This glycoprotein occurs in extracellular matrix and was named because of its association with fibroblasts and binding to fibrin during blood clotting (Yamada et al, 1974). Two types of fibronectins are recognized. Cell surface or cellular fibronectin is synthesized by a variety of cells: fibroblasts, myoblast, and certain epithelial cells (Yamada et al, 1978). It has molecular weight of 220,000-240,000 daltons and contains 5% carbohydrate. Plasma fibronectin, "cold insoluble globulin" has a slightly smaller weight (200000-220000) an it precipitates in the cold if complexed to fibrinogen. Plasma fibronectin is a dimer of two disulfuride bonded polypeptide (subunits) whereas cellular fibronectin occurs as dimers and monomers. Most antisera to the two forms of fibronectin cross react (Vaheri et al, 1978; Yamada et al, 1978; Yamada et al 1974) but a monoclonal antibody has been described that distinguishes between the two (Atherton et al, 1981). Fibronectin has a biological role in linking cells, collagen, and fibrin (Yamada et al, 1974; 1978). Fibronectin together with laminin and collagen appear in the mouse embryo at the morula and blastocyte stage as shown by Hay (1981) and it has been detected by immunofluorescence in basement membranes (Stenman et al, 1978; Lindner et al,1975) and cell membranes (Gay et al, 1978).

Laminin: This largely glycoprotein has been detected in basement membranes (basal laminae) immunofluorescence (Timpl et al 1979) and by immunoelectron microscopy (Madri et al, 1980). Laminin is composed of at least two polypeptide chains, one 220000 in molecular weight, the other 440000 joined by disulfuride bonds. It is relatively insoluble and difficult to characterize but is clearly involved in epithelial attachment to collagen (Terranova et al, 1980).

Comment: Extracellular matrices are composed of macromolecule that are unique to each tissue and are capable of influencing the function of cells within the tissues. Matrices provide not only structural support to the tissues, but also filter the substances that reach cells and can affect what leaves the body as demonstrated by Hay (1981). Matrix components also have biological activities such as promoting cell growth, differentiation, migration, adhesion and transformation. Cell collagen interactions of cells with their matrix could result in tissue malfunction: poor wound healing, abnormal embryological development, enhanced tissue aging, fibrotic disease and metastatic spread of tumor cells. (Kleinman et al, 1976).

Developmental and age-associated changes in the human structural proteins in the cornea: Development and aging

Collagens, fibronectin and laminin play a major role in the structure of the human cornea as in various other tissues (Kleinman et al, 1976; 1981). These glycoproteins are constituents mainly of the basements membranes of the epithelium and Descemet's membrane (Fujikawa et al, 1981; 1984). They are synthesized by corneal endothelial cells and play a role in the wound healing process (Fujikawa et al, 1981; 1984; Meier et al, 1974).

Type I collagen: Initial biochemical evidence for a possible heterogeneity of collagen types in the avian eye was obtained by Trelstad and Kang (Trelstal et al, 1974) who analyzed cornea, lens, vitreous body, and sclera from 1500 chick eyes. They described type I collagen in the cornea and fibrous sclera while the lens capsule and vitreous body suggested the presence of a different collagen. Type I collagen was found later to constitute the major collagen type in the cornea of other vertebrates. (Kleinman et al 1981; Wartiovaara et al, 1980; Yanoff et al, 1975).

Type II collagen: In the chick cornea, immunofluorescence

studies (Von der Mark et al, 1972) demonstrated that Type II collagen appears the primary stroma at stage 20 (3 days of gestation) and persists until stage 35 (14 days) in the anterior part of the secondary stroma as well as in Descemet's membrane (Hay et al, 1979). It is not found in the cornea of adult chick. Antibodies to type IV collagen stain Descemet's membrane and Bowman's membrane in the cornea of chick, mouse, calf, and human: this also appears to be the only collagen present in the lens capsule (Kleinman et al, 1981). Biochemical and immunohistological studies on the collagen in chick vitreous humor suggests that it is related to type II collagen (Trelstad et al, 1974; Ruoslahti et al, 1978; Hay et al, 1978; Swan et al, 1980) although it shows an electrophoretic mobility slightly different from 1(II) obtained from articular cartilage (Swan et al, 1980). Type II was not shown to be present in the adult cornea but embryonic chick corneal epithelium synthesizes type I and II collagens (Lindenmeyer et al, 1977).

Type III collagen has been reported to appear in the mouse embryo at the same developmental stages and the same site as type I collagen (Leivo et al, 1980). Type I, and IV collagens were detected in human keratoconus cornea and found in similar distribution to that of normal cornea. Type III was found only in the scar (Newsome et al, 1981). The authors revealed significant type IV collagen in the central cornea.

Type IV collagen: Monoclonal antibodies against chicken type IV collagen revealed that the basement membrane of the corneal epithelium showed little, if any, staining (Fitch et al, 1982). On the other hand monoclonal antibody to human basement membrane Type IV collagen shows intense reactivity of basement membranes from a variety of human tissues but does not stain tissues of bovine, rabbit, rat or mouse origin (Sakai et al, 1982). These findings were confirmed by others (Wulle et al, 1969) who developed a hybridoma cell line that synthesizes antibody to human basement membrane Type IV collagen (Sundarray et al, 1982). This monoclonal antibody does not cross-react with other types of collagen including Type I, III and V as determined by ELISA (enzyme-linked immunoabsorbant assay) and by immununohistochemical staining of corneal and lens tissues. Descemet's membrane of mouse, rabbit, and human corneal epithelium, and lens capsule are both abundant sources of type IV collagen (Sundarray et al, 1982). Recent research showed that the extracellular matrix of adult vertebrate corneal stroma is composed primarily of the interstitial collagen type I and smaller amounts of type III and type IV collagen (Pratt et al, 1985). These collagens are organized into overlapping lamellae of striated filaments. In addition to these lamellar structures, the cornea stroma also contains 100 to 250 nm bundles of nonstraited 8 to 22 nm microfibrils by immunofluores cent and electron microscopic immunolocalization. These microfibrils bundles in the mouse are associated with type III collagen, type IV collagen, elastin or oxytalan microfibrils. An *in vitro* study showed that type IV collagen tends to form distinct complexes within laminin (Pratt et al, 1985).

Fibronectin: The distribution of fibronectin in developing rabbit cornea by immunohistofluorescent staining of cryostat sections of cornea from 13,15 and 20 day old fetuses, 3 day postpartum and adults was recently described (Cintron et al, 1984). At 15 days of gestation, fluorescence was associated with the stromal extracellular matrix of the cornea, the subepithelial zone and the lens capsule. In the 20-day fetus an intense fibronectin fluorescence was present along with the inner corneal stromal border coincident with the formation of

Descemet's membrane. Fluorescence within the corneal stroma appeared as fines lines restricted to the collagen lamellae, remaining through birth and disappearing in the adult. Although stromal fluorescence disappeared in the adult, Descemet's membrane continued to display fluorescence but to a lesser extent (Cintron et al, 1984) Fibronectin and laminin in normal human cornea form donors of various ages revealed that fibronectin was present in the epithelial basement membrane of fetal (20 weeks) and adult cornea but was absent in the corneas from postnatal donors. Fibronectin also appeared in Descemet's membrane, diminishing in the corneas of elderly individuals.

Fibronectin was present as diffuse fine fibers in the fetal (20 weeks) and adult cornea stroma and was diminished in the corneal stroma from other age groups. Fibronectin was present in endothelial basement membrane of the blood vessels at the corneal limbus (Moro et al, 1977). Laminin (LN) and type IV collagen were present in the epithelial basement membrane as delicate linear bands at the limbus, in Descemet's membrane as double linear bands and in the stroma as scattered, large fibers. They were also present in the endothelial basement membrane of limbal blood vessels. Laminin was found in the epithelial basement membrane of the central cornea, while type IV collagen was not. (Fujikawa et al, 1985). During mouse embryogenesis, laminin is the first compound of extracellular matrix that is detected (Leivo et al,1980). Type IV collagen, another basement membrane component becomes detectable later, and both are present in all embryonic and extraembryonic basement membrane. Some tumors synthesize laminin (Strickland et al, 1980) upon differentiation; thus laminin can be used as a marker for studies on differentiation, and embryogenesis (Reddi et al, 1984).

Bullous pemphigoid antigen (BPA): Frozen sections of rabbit cornea were incubated with pemphigoid serum and then with fluorescein -conjugated goat antihuman IgG to detect BPA, a component of the lamina lucida. Results showed that BPA antigen is present in the normal epithelial basement membrane of rabbit, but is absent on the wounded surface. It appears in a healing cornea at 2 weeks coinciding with the time fibronectin is diminishing (Fujikawa et al, 1981). In human corneas of various age groups, BPA was present in epithelial basement membrane of all corneas studied.

Development and age-associated changes in human cornea

In our study, human fetal and selected early postnatal and adult corneas were collected. Donor eyes ranging from 8 weeks, 9 weeks, 10 weeks, 11 weeks, 13 weeks, 15 weeks, 17 weeks, 20 weeks, 27 weeks, 36 weeks, and 38 weeks of gestation, 4 months, and 9 months post-natal and normal adult (11 years, 48 and 77 years old) were compared. Polyclonal antibodies to type I, type II, Type III, and type IV were used. Fibronectin and laminin were obtained as described (Newsome et al, 1981). The fluorescent antibody technique was applied (Newsome et al, 1981) to determine the exact distribution and localization of each type of collagen in the corneas of various age groups. Direct immunofluorescent staining was used for fibronectin. Our results showed as follows:

Type I collagen was detected at the limbus and the corneal stroma as early as 8 weeks of gestation (Fig. 1A). At 11 weeks of gestation it was found at the epithelial basement membrane (EMB) and in Descemet's membrane (DM) in addition to the stroma (Fig. 1B). Immunofluorescence was more marked in DM posteriorly. The findings apply throughout the fetal and adult corneas at the various ages up to old age (77 years old) (Fig. 1C,D).

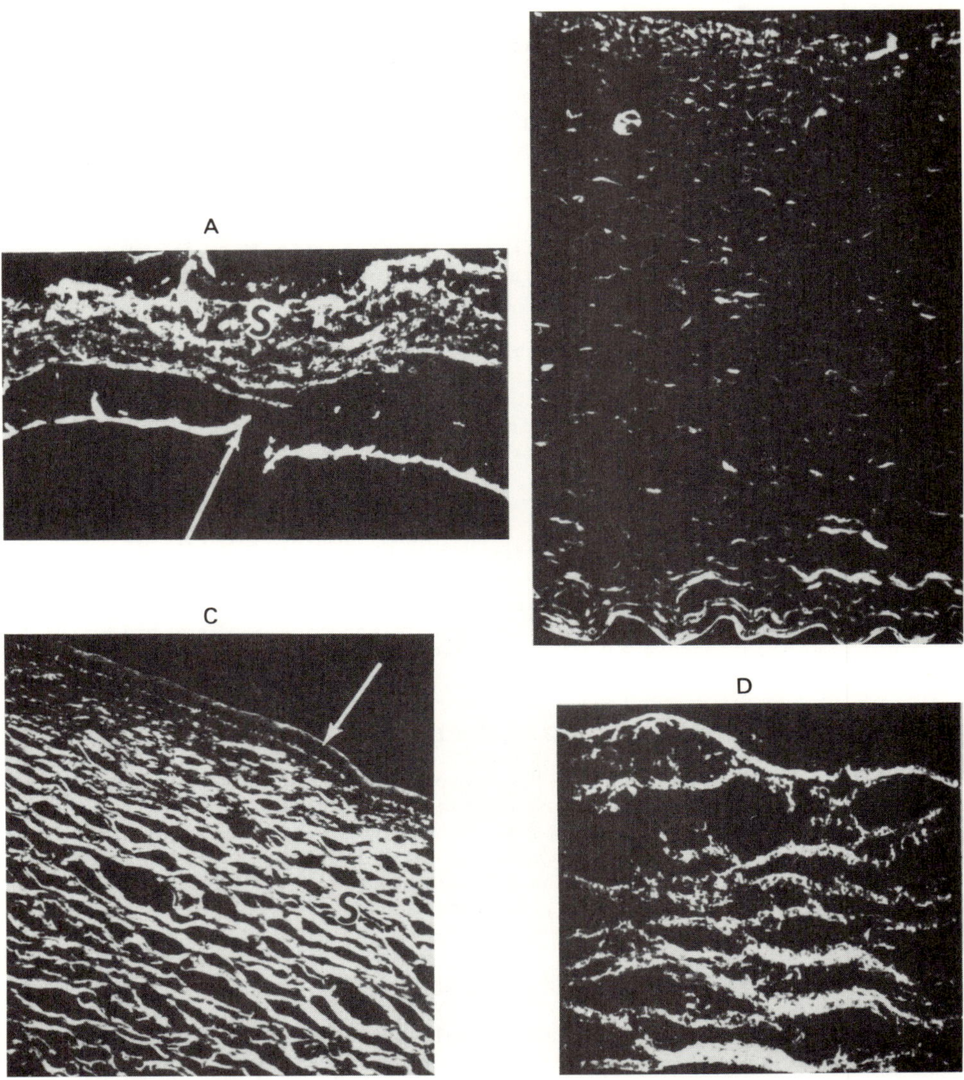

Fig.1. Immunofluorescence (IF) with antibodies to type I Collagen.

A. 9 weeks gestation. Diffuse staining of corneal stroma(S). Lens capsule is indicated (arrow)(X63).

B. 10 1/2 weeks gestation. Stromal staining is more marked posteriorly. Epithelial basement membrane is indicated arrow (X100).

C. 36 weeks gestation. Diffuse staining of the stroma(S). epithelium is indicated (arrow)(X40).

D. Normal human cornea staining of stroma,Bowman's layer and faint staining of Descemet's membrane (X40).

<u>Type II collagen:</u> Polyclonal antibody to type II reveal staining for this component in the human cornea at any stage of development and aging.

A

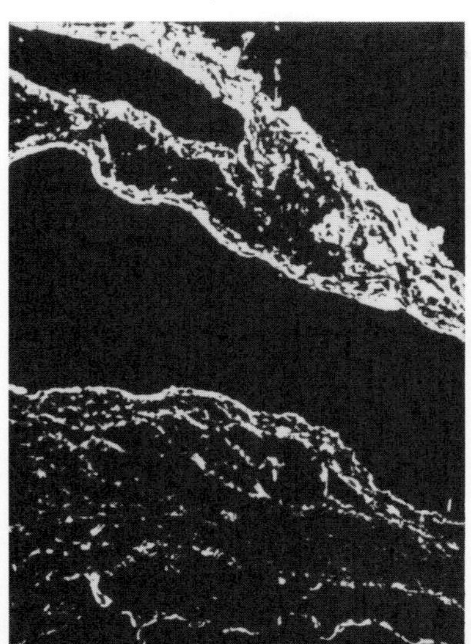

B

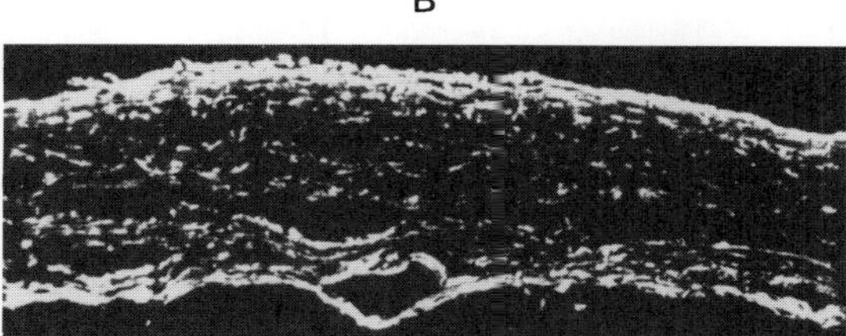

C

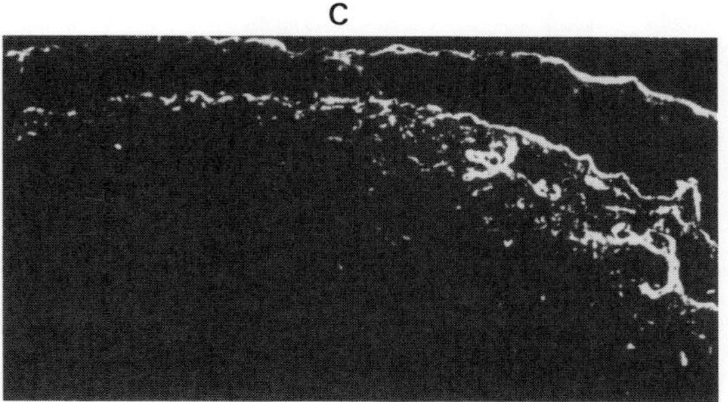

Fig. 2. Immunofluorescence with antibodies to type III collagen.
A. 10 1/2 weeks gestation. Staining of corneal and scleral collagen (X40)
B. 17 weeks gestation.Diffuse staining of corneal stroma, epithelial basement membrane and Descemet's membrane (X200).
C. Normal human cornea. Positive staining only in the limbus (arrow). Epithelium is indicated (E)(X40).

<u>Type III collagen:</u> Type III collagen was found in early fetal life (8-20 weeks of gestation) (Fig. 2A,B) in all the corneas that were observed. After 27 weeks of gestation, type III collagen could not be detected in the corneal stroma of fetal eyes or in the adult normal human cornea (NHC), except at the limbus (Fig.2C).

A

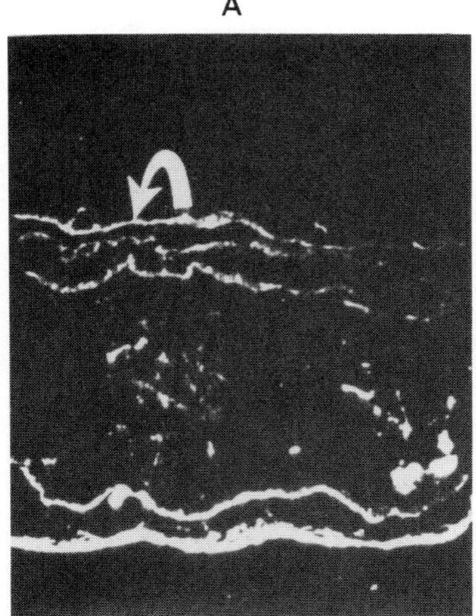

B

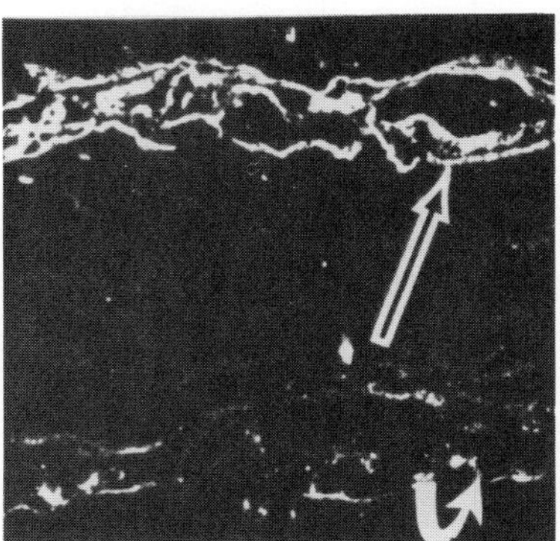

C

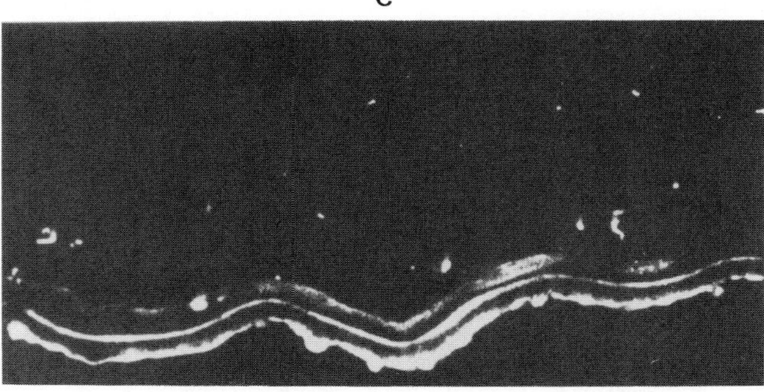

Fig. 3 Immunofluorescence with antibodies to type IV collagen.
A. 8 week gestation. Staining of corneal epithelial basement membrane (EB) (arrow) (X63).
B. 9 week gestation. Staining of EBM of conjunctiva and cornea (arrow). Patchy staining of Descemet's membrane (curved arrow) (X63).
C. Normal human cornea.(48 year old). Double linear staining of Descemet's membrane. (X63).

Type IV collagen: Type IV collagen was detected in the EBM as early as 8 weeks of gestation (Fig. 3A) and remained positive through out fetal and adult life. Descemet's membrane was negative at 8 weeks of gestation but was focally present at 9 weeks of gestation (Fig. 3B), and was well defined at 11 weeks of gestation. EBM and DM remained positive throughout fetal and adult life but the solid line of DM in the fetus became a double line postnatally (Fig. 3C). In the adult cornea, the central EBM showed only faint staining compared to the staining at the limbus.

A

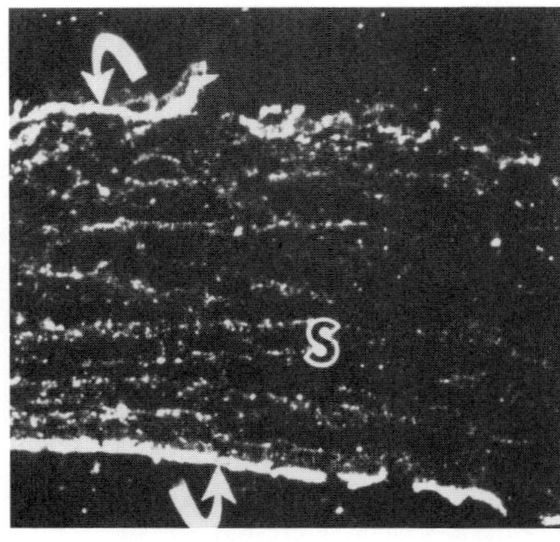

B

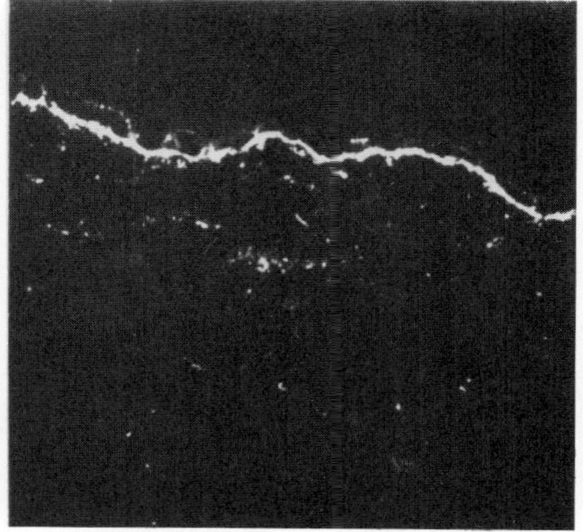

C

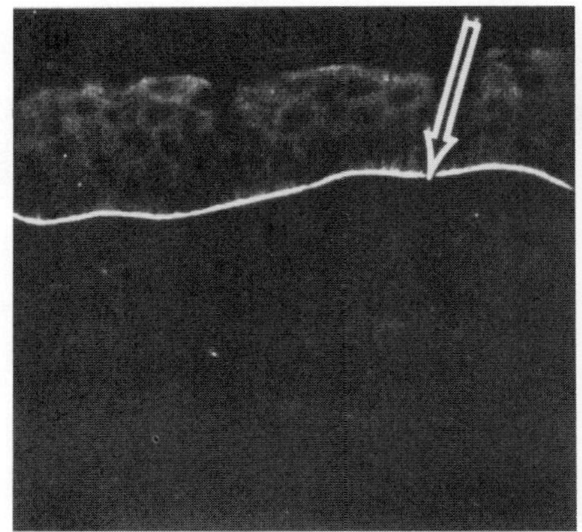

D

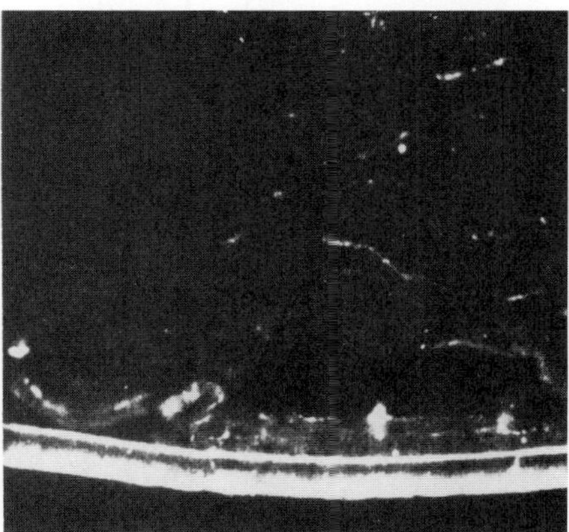

Fig. 4. Immunofluorescence with antibodies to fibronectin.

A. 10 1/2 weeks gestation. Staining of EBM (curved arrow), stroma (S) and Descemet's membrane (lower arrow)(X63).

B. 15 weeks gestation. Staining of epithelial basement membrane. Faint staining of stromal fibrils was also present (X100).

C. Adult (66 years old) human cornea.Staining of epithelial basement membrane (X200).

D. Adult (66 years old) human cornea.Faint staining of Descemet's membrane (X200).

<u>Fibronectin</u> (FN): Staining of FNH of DM was negative at the very early age of 8-9 weeks gestation (Fig. 4A), became positive at 11 weeks of gestation and remained positive throughout fetal (Fig. 4B) and adult life (Fig. 4C). The cornea stroma and EBM showed some staining from early fetal life up to the age of 30 weeks of gestation. They became negative for FN at the age of 38 weeks of gestation and remained so throughout adulthood up to old age and then became positive again. In corneas from old donors, fluorescence for FN in DM seemed to lessen (Fig. 4D).

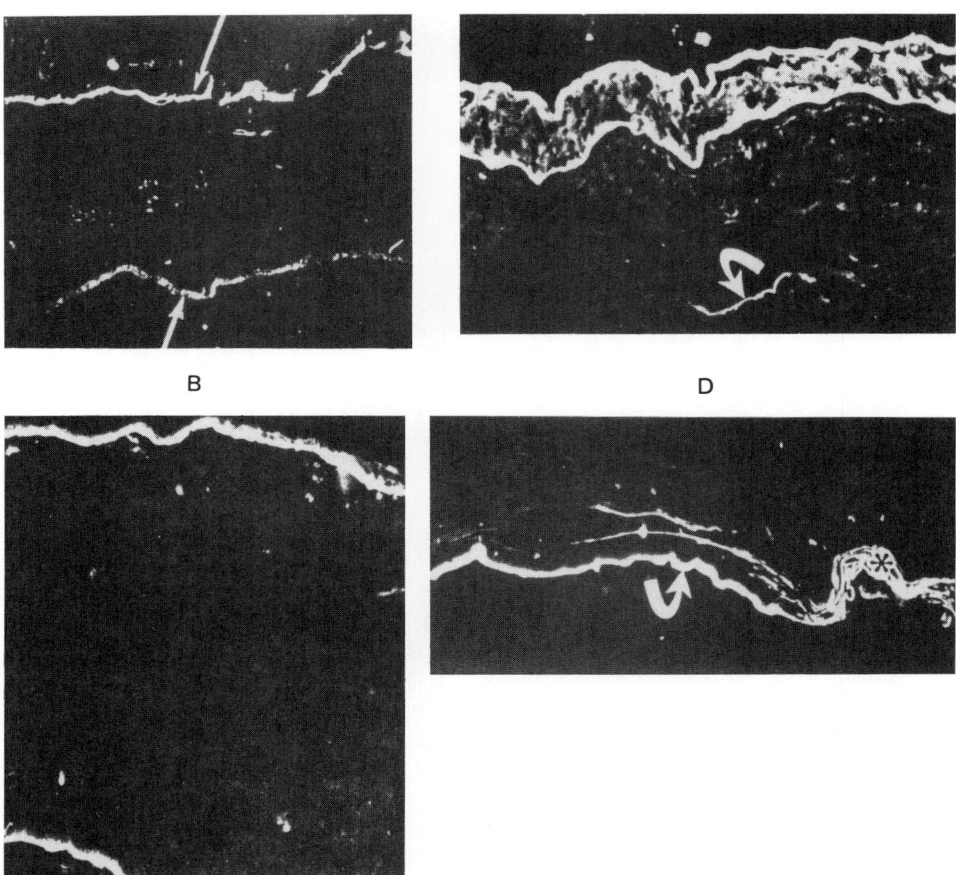

Fig. 5. Immunofluorescence with antibodies to laminin
A. 9 weeks gestation. Staining of the epithelial basement membrane (upper arrow) and Descemet's membrane (lower arrow).
B. 15 weeks gestation. More pronounced staining of epithelial basement membrane and Descemet's membrane (X63).
C. Adult human cornea (77 years old). Staining of epithelial basement membrane and occasional stromal fibers (arrows) (X63).
D. Adult human cornea (77 years old). Staining of Descemet's membrane with Hassall Henle bodies(arrow) and anterior trabecular meshwork (*)(X63).

<u>Laminin</u> (LN) staining for laminin was positive in the EBM from the age of 8 weeks gestation throughout fetal and adult life (Figs. 5A-D). At the age of 8 weeks of gestation the staining for laminin in DM was negative,but became positive at the age of 9 weeks of gesta tion, before the morphological appearance of DM. Positive staining for laminin continued throughout fetal and adult life where it appears as a single band at the fetal period and changes to a double line postnatally, becoming a single line again on the posterior aspect of DM in the elderly,(Hassall Henle bodies)(Fig. 5D).

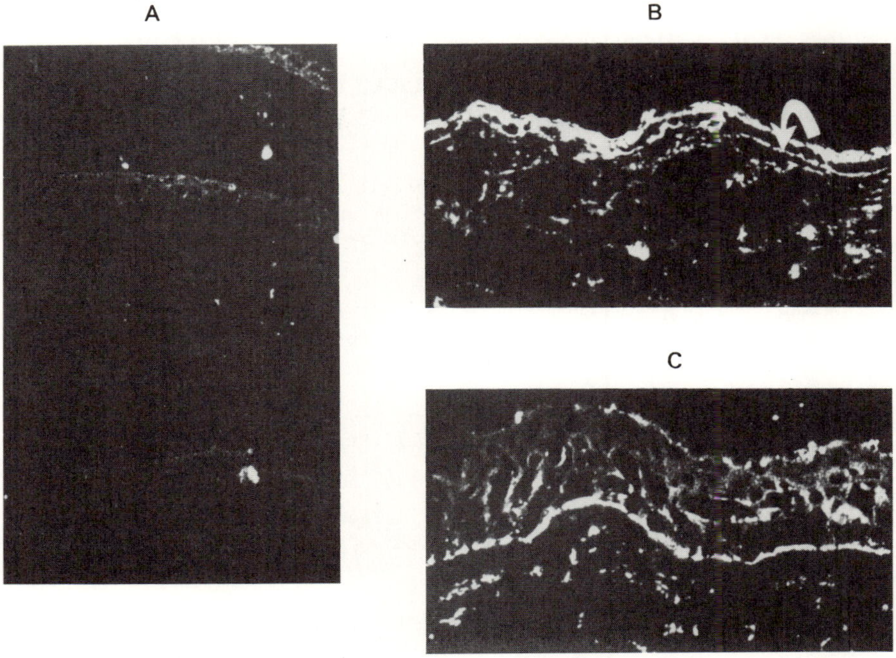

Fig. 6. Immunofluorescence with antibodies to bullous pemphigoid antigen

A. 10 1/2 weeks gestation. Negative stains of conjunctiva and corneal epithelia(X100).

B. 16 week gestation. Patchy staining of central corneal epithelial basement membrane (arrow) (X100).

C. 4 months postgestation. Faint stain of epithelial basement membrane.

Bullous pemphigoid antigen (BPA): Staining with polyclonal antibody for BPA did not reveal any of this antigen at the early age of gestation (8-10 1/2 weeks)(Fig. 6A) and traces in the EBM at the age of 15-17 weeks of gestation (Fig.6B). It became more positive after 20 weeks of gestation and remained so in the EBM throughout the fetal and adult life (Fig. 6B).

Perspectives

The aging and developmental changes in the cornea starts at the moment of conception, and continues throughout the entire fetal and adult life. The aging and developmental process in the cells and the structural proteins of the corneal matrix have been briefly discussed, emphasizing the use of new polyclonal and monoclonal antibodies as a tool for the evaluation of embryogenesis and tissue specificity. The basic knowledge of these age related changes may lead to a better understanding of healthy and diseased states.

References

1. Andrew, J.S. and Kuwabara T. 1961, Net triglyceride synthesis by rabbit cornea in vitro, Biochem Biophys 54:315-321.

2. Atherton, B.T. and Hymes R.O., 1981, A difference between plasma cellular fibronectins located with monoclonal antibod ies, Cell 25:133-141.

3. Bentz, H., Bachinger, H.P., Glanville R. and Kuhn,K.,1978,Phys ical evidence for the assembly of A and B chains of human placenta collagen in single triple helix. Eur. J. Biochem. 95:563-567.

4. Berman, M., Manseau, E., Low, M., Aiken, D., 1983, Ulceration is correlated with degradation of fibillin and fibronectin at the corneal surface, Invest. Ophthalmol. Vis. Sci. 24: 1358-1366.

5. Bornstein, P., and H. Sage., 1980, Structurally distinct collagen types, Ann Rev. Biochem. 49:954-1003.

6. Bourne, W.M., and Kaufman, H.E., 1984, Specular microscopy of human corneal endothelium in vivo, Am. J. Ophthalmol 81:319-323.

7. Cintron, C.,Fujikawa, L.S., Covington, H., Foster, C.S., Colvin, R.B., 1984, Fibronectin in developing rabbit cornea, Curr. Eye Res. 3:489-499.

8. Chi, H.H., Teng, C.C., and H.M. Katzin, 1958, Histopathology of primary endothelial-epithelial dystrophy of the cornea, Am. J. Opthalmol 45:518-535.

9. Cogan, D.G., and Kuwabara, T., 1958, Lipid keratopathy and atheroma. Circulation 18:519-535.

10. Cogan, D.G., and Kuwabara, T., 1959, Arcus senilis, Arch Ophthalmol. 61:553-560.

11. Cogan, D.G., and Kuwabara, T., 1959, Ocular changes in experimental hypercholesteremia, Arch. Ophthalmol. 61:219-228.

12. Cogan, D.G., and Kuwabara, T., 1971, Growth and regenerating potential of Descemet's membrane, Trans. Opthalmol. Society U.K. 91:875-894.

13. Dodson, J.W., and Hay, E.D., 1974, Secretion of collagen by corneal epithelium, J. Exp.Zool. 189:51-72.

14. Eyre, D.R., 1980, Collagen: molecular diversity in the body's protein scaffold, Science 207: 1315-1322.

15. Fitch, M.M., Gibney, E., Sanderson, D., Mayne, R., and Linsenmeyer, Production and of monoclonal antibody to human type IV collagen. Am. J. Path. 108: 310-318.

16. Foidart, J.M., and Reddi, A.H., 1980, Immunofluorescent localization of type IV collagen and laminin during endochon dral bone differentiation and regulation by pituitary growth hormone, Dev. Biol. 75:130-136.

17. Foidart, J.M., Bere, E.W., Yaat, M., Martin, G.R. and Katz, S.K., 1980, Distribution and immunoelectron microscopic localization of laminin, a noncollagenous basement membrane glycoprotein, Lab Invest., 42: 336-342.

18. Fujikawa, L.S., Foster, C.S., Harrist, T.J., Lanigan, J.M., and Colvin, R.B., (1981), Fibronectin in healing rabbit corneal wounds, Lab Invest 45:120-129.

19. Fujikawa, L.S., Foster, C.S., Gipson, I.K., Colvin, R.B., (1981), Fibronectin in healing rabbit corneal wounds, Lab Invest 45:120-129.

20. Fujikawa, L.S., Lee, R.E., Guillaumin, E.A., Vastin, D.W., Foster, C.S., and Colvin, R.B.,1985, Invest Opthalmol. Vis. Sci. In press.

21. Gay, S., Miller, E.J., 1978, Physiology and pathology of connective tissue. Academic Press,Inc., New York, 334-345.

22. Hay, E.D., Revel, J.P., 1969, Fine structures of the developing avian cornea, Monograms in developing Biology, Vl Karger, A.G., Basal, S. 23. Hay, E.D., 1981, Extracellular matrix, J. Cell Biol. 91:205-233.

23. Hay, E.D., 1981, Extracellular matrix, J. Cell Biol. 91:205-233.

24. Hay, E.D.,Linsenmayer, T.F., Trelstad, R.L.,and Von der Mark, K. 1979, Origin and distribution of collagens in the developing avian cornea, Curr. Top Eye Res. 1:1-5.

25. Hogan, M.J., Alvarado, J.A., and Wedell, J.E., 1962 Histology of the human eye. S.W. Saunders,Philadelphia, p. 277-344.

26. Hogan, M.J., Zimmerman, L.E., 1961, The cornea and sclera, Ophthalmic Pathology, S.W. Saunders, Philadelphia, pp. 277-344.

27. Johnston, M.C., Bhakdinarouk, A.,Reid, Y.C., 1974, An expanded role of the neural crest in oral and pharyngeal development, in Fourth Sympsoium on Oral Sensation and Perception, Washington, D.C.,pp.37-52.

28. Kanai, A.,Wood, T.C.., Polack, F.M., and Kaufman, H.E., 1971,The fine structure of sclerocornea. Invest. Ophthalmol. Vis Sci. 10:687-694.

29. Keihl, F.A., and Darnell, C.A., 1939, Replication of Descemet's membrane, Arch. Ophthalmol. 22:672-674.

30. Kleinman, H.K., Klebe, R., and Martin, G.R., 1981, Role of collagenous matrixes in the adhesion and growth of cells. J.Cell Biol. 88:473-485.

31. Kleinman, H.K., Rohrback, D.H., Terraova, V.P., Varner, H.H., Hewitt's, A.T., Martin, G.R., and Schiffman, E., 1982, Collagenous matrices as determinants of cell function, in Immunochemistry of the extracellular matrix, Furthmayr,H., CRC Press, Florida, pp. 151-174.

32. Kuwabara, T., 1976, Effect of electronic strobe flashlight on the monkey retina. Jpn J. Ophthalmol 20:9-18.

33. Kuwabara, T., 1979, Age related changes of the eye in Special senses in aging. Seong, S., Hanandorothy, H. Coon. The University of Michigan, pp. 46-78.

34. T. Kuwabara, 1978, Current concepts in anatomy and histology of the cornea.. Contact Ocular Lens Med.J. 4:101-132.

35. Kefalides, N.A., 1978, Current status of chemistry and structure of basement membranes in Biology and chemistry of basement membranes. Nicholas A. Kefalides, Academic Press, Inc. New York, pp. 215-228.

36. Laing R.A., Sandstorm M.M., and Leibowitz, H.M.,1975, In vivo photomicroscopy of the corneal endothelium. Arch. Opthalmol 93;143-145.

37. Linder E., Vaheri A., Ruoslahti E., and Wartiovaara J., 1975, Distribution of fibroblast surface antigen in the developing chick embryo. J. Exp. Med. 142:41-49.

38. Linsenmayer, T.F., Smith, G.N., Hay E.D., 1977, Synthesis of two collagen types by embryonic chick embryo in vitro., Proc. Natl.Acad. Sci. USA 74:39-43.

39. Leivo, I., Vaheri, A, Timpl, R. and Wartiovaara J., Appearance and distribution of collagen and laminin in the early mouse embryo, Dev. Biol. 76:100-114.

40. Mayne, R., Vail, M.S. and Miller, E.J., 1980, Analysis of changes in collagen biosynthesis that occur when chick chrondrocytes are grown in S-bromo-2- deoxyuridine, Acad. Sci. U.S.A. 72:4511-4515.

41. Mayne, R., Vail, M.S. Mayne, P.M., and Miller E.J.,1975, Changes in types of collagen synthesized as clones of chick chondrocyte grow and gradually lose division capacity, Proc. Natl. Acad. Sci. U.S.A. 73:1674-1678.

42. Madri,J.A., Roll F.J., and Foidart, J.M., 1980. Ultrastructural localization of fibronectin and laminin in the basement membranes on the murine kidney., J. Cell Biol. 86:682-687

43. Miller, E.J. and Matukas, V.J., 1974. Biosynthesis of collagen, biochemist's view. Fed Proc 33:1197-1204.

44. Moro, L. and Smith B.D., 1977. Identification of collagen a-1 (1) trimer and normal type 1 collagen in a polyoma virus-induced mouse tumor Arch Biochem. Biophys. 182: 33-41.

45. Meier, S. Hay, E.D., 1974. Control of corneal differentiation by extracellular materials. Collagen as a promoter and stabilizer of epineural stroma production. Dev. Biol 38:249-270.

46. Newsome, D.A., Foidart, J.M. Hassell, J.R.,Krachmer, J.H., Rodrigues, M.M. and Katz, S.I., 1981. Detection of specific collagen types in normal and keratoconus corneas.Invest. Ophthalmol. Vis Sci. 20:783-750.

47. Pei, Y.F., Rhodin, J.A., 1971. Electron microscopic study of the development of the mouse corneal epithelium. Invest. Ophthalmol Vis. Sci. 11:822-835.

48. Pratt, B.M., Madri, J.A., 1985. Immunolocalization of type IV collagen, laminin, in nonbasement membrane structures of murine corneal stroma, Lab. Invest. 52:650-656.

50. Ruoslahti, E., Engval, E., 1978. Immunochemical and collagen-binding properties of fibronectin Ann. NY Acad. Sci. 312-178-191.

51. Sakai, L.Y., Engrall, E., Hollister, D.W. Burgeson,R., 1982. Production and characterization of monoclonal antibody to human type IV collagen. Am.J.Pathol 108:310-318.

52. Smelser, G.K. and Ozanics, V., 1972. Development of the cornea in Symposium on the cornea. New Orleans Academy of Ophthal mology, C.V. Mosby.

53. Stenman, S. and Vaheri A., 1978. Distribution of a major connective tissue protein, fibronectin, in normal human tissues. J. Exp. Med. 147:1054-1064

54. Strickland, S., Smith, K.K., Marotti, K.M., 1980. Hormonal induction of a differentiation in teratocarcinoma stem cells: generation of parietal endoderm by retinal acid and dibutyryl camp. Cell 21:347- 355.

55. Sundarraj, N., Wilson, J., 1982. Monoclonal antibody to human basement membrane collagen type IV. Immunology 47:133-140.

56. Swann, D.A., and Sotman, S.S., 1980. The chemical composition of bovine vitreous-humor collagen fibers. Biochem. J. 185:545-549.

57. Terranova, V.P., Rohrbach, D.H. and Martin, G.R.,1980. Role of laminin in the attachment of Pam 212 (epithelial) cells to basement membrane collagen. Cell 22:719-726.

58. Timpl, R., Rohde, H., Gehron, P. and Martin, G.R.,1979. Laminin: a glycoprotein from basement membranes. J. Bio Chem. 254:9933-9937.

59. Trelsted, R., Coulombre, A., 1971. Morphogenesis of the collagenous stroma in the chick cornea. J.Cell Biol. 50:840-853.

60. Trelstad, R., 1970. The golgi apparatus in chick corneal epithelium: changes in intracellular position during develop ment. J. Cell Biol. 45:34-42.

61. Trelstad, R.L., and Kang, A., 1974. Collagen heterogeneity in the human eye: lens, vitreous body, cornea and sclera. Exp. Eye Res. 18:395-397.

62. Tschetter, R.T., 1966. Ciliary body cysts in multiple myeloma. Arch Ophthalmol. 73:686-699.

63. Vaheri, E. and Mosher, D.F., 1978. High molecular weight, cell surface-associated glycoprotein (fibronectin)lost in malignant

transformation. Biochem. Biophys. Acta 516:1-25.

64. Von der Mark, K., Von der Mark, H., Timpl, R. and Trelstad, R.I., 1972. Immunofluorescent localization of collagen types I,II, and III in the embryonic chick egg. Dev. Biol. 59:75-78.

65. Wartiovaara, J., Leivo, I. and Vaheri, A.,1980. Matrix glycoprotein in the early mouse development and in differentia tion of teratocarcinoma cells in The cell surface mediator of developmental processes. (ed) S. Subtelny and Wessels, N.K.., pp. 305-324. Academic Press, New York.

66. Wulle, K.G. and Leache, W., 1969. Electron microscopic observations of the early develoment of the human corneal endothelium and Descemet's membrane Opthalmologica 157:451-461.

67. Wulle, K.G., Runrecht, K.W. and Windrath, L.C.,1974. Electron microscopy of the development of cell junctions in the embryonic and fetal human corneal endothelium Invest Ophthalmol Vis. Sci. 23:923-924.

68. Wulle, K.G., 1972. Electron microscopy of the fetal development of the corneal endothelium and Descemet's membrane of the human eye. Invest. Ophthalmol. Vis.Sci. 11-892-904.

69. Yamada, K.M., and Olden, K.,1978. Fibronectins: Adhesive glycoproteins of cell surface and flood. Nature (London) 276:179-184.

70. Yamada, K.M. and Weston, J., 1974. Isolation of a major cell surface glycoprotein from fibroblasts. Proc. Natl. Acad. Sci. U.S.A. 3492-3496.

71. Yanoff, M. and Fine, B.S., 1975. Ocular Pathology Hagerstown, Harper and Row, pp. 305-398.

72. Zinn, K.M., 1970. Changes in corneal ultrastructure resulting from early lens removal in the developing chick embryo. Invest. Ophthalmol. Vis. Sci. 9:165-182.

UPDATE 1990

BIBLIOGRAPHY

1. Ben-Zvi, A., Rodrigues, M.M., Krachmer, J.H. and Fujikawa, L., 1986, Collagen staining in corneal tissues. Curr. Eye Res. 5:105-117.

2. Yue, B.Y.J.T., Sugar, J., and Schrode, K., 1986, Collagen staining in corneal tissues. Eye Res. 5:559-564.

3. Nakayasu, K., Janaka, M., Konomi, H. and Hayashi, T., 1986, Distribution of types I, II, III, IV, and V collagen in normal and keratoconus corneas. Ophthal. Res. 18:1-10.

4. Zimmerman, D.R., Trueb, B. Winterhalter, K.H., Witmer, R. and Fischer, R.W., 1986, Type VI collagen is a major component of the human cornea. FEBS Lett. 197:55-58.

5. Rodrigues, M., Ben-Zvi, A., Krachmer, J., Schermer, A. and Sun, T.T., 1987, Suprabasal expression of a 64- kilodalton keratin (No. 3) in developing human corneal epithelium. Differentiation 34:60-67.

6. Rodrigues, M.M., Krachmer, J., Rajagopalan, S. and Ben-Zvi, A., 1987, Actin filament localization in developing and pathologic human corneas. Cornea 6:190-196.

7. Tisdale, A.S., Spurr-Michaud, S.J., Rodrigues, M., Hackett, J., Krachmer, J., Gipson, I.K., 1988, Development of the anchoring structures of the epithelium in rabbit and human fetal corneas. Ophthalmol. Vis. Sci. 29:727-736.

8. Zimmerman, D.R., Fischer, R.W., Winterhalter, K.H., Witmer, R. and Vaughan, L., 1988, Comparative studies of collagens in normal and keratoconus corneas. Exp. Eye Res. 46:431-442.

9. Hirano, K., Kobayashi, M., Kobayashi, K., Hoshino, T. and Awaya, S., 1989, Experimental formation of 100 nm periodic fibrils in the mouse corneal stroma and trabecular meshwork. Ophthalmol. Vis. Sci. 30:869-874.

10. Morton, K. Hutchinson, C., Jeanny, J.C., Karpouzas, I., Pouliquen, I.Y., Courtois, Y., 1989, Colocalization of fibroblast growth factor binding sites with extracellular matrix components in normal and keratoconus corneas. Curr. Eye Res. 8:975-987.

11. Sevel, D. and Isaacs, R., 1989, A re-evaluation of corneal development. Trans. Am. Ophthalmol. Soc. 86:178-204.

ROLE OF ALDOSE REDUCTASE AND EFFECTS OF ALDOSE REDUCTASE INHIBITORS IN OCULAR TISSUE AGING PHENOMENA IN THE DIABETIC RAT

Brenda W. Griffin, M. L. Chandler, Louis DeSantis, and B. M. York

Alcon Laboratories, Inc.
P. O. Box 6600, 6201 South Freeway
Fort Worth, Texas 76115

INTRODUCTION

Studies of the role of aldose reductase in sugar cataractogenesis have been aided by the synthesis and characterization of inhibitors of this enzyme which can prevent lens opacification in galactosemic or diabetic rats[1,2]. The anti-cataract activity of aldose reductase inhibitors correlates with inhibition of polyol accumulation in the lenses of treated galactosemic or diabetic animals; in untreated diabetic rats, lens sorbitol levels may be elevated 100-fold over the normal value[3]. Hyperglycemia produces an increase in glucose content of tissues not responsive to insulin. Due to increased substrate availability, the rate of sorbitol formation by action of aldose reductase is greatly accelerated relative to conversion of this metabolite to fructose via sorbitol dehydrogenase[1]:

$$\text{Glucose} + \text{NADPH} + H^+ \xrightarrow{\text{Aldose Reductase}} \text{Sorbitol} + \text{NADP}^+ \quad (1a)$$

$$\text{Sorbitol} + \text{NAD}^+ \xrightarrow{\text{Sorbitol Dehydrogenase}} \text{Fructose} + \text{NADH} + H^+ . \quad (1b)$$

Since polyols such as sorbitol are poorly transported across cell membranes, an early effect of intracellular sorbitol accumulation appears to be increased osmolality[1]. Compensatory responses of cells to an osmotic insult include water uptake and swelling, which, if sustained chronically, produce membrane damage and other alterations that progress eventually to end-stage nuclear cataract[1,2]. The demonstration that aldose reductase inhibitors can prevent cataract in uncontrolled diabetic rats, without affecting the hyperglycemic status of the animals, argues aginst a direct cataractogenic effect of glucose[3,4].

Aldose reductase inhibitors have also been shown to have beneficial effects on other tissues of diabetic rats, i.e., peripheral nerve, kidney, and retina, which display pathophysiological changes very similar to those of the human disease[5,6]. Diabetic retinopathy appears to be one of the serious consequences of a generalized microangiopathy[7]. The morphological features of human diabetic retinopathy include an early, selective loss of capillary mural cells (pericytes) and later capillary basement membrane thickening[5-6]. It was shown by immunochemical techniques that pericytes in human retinal capillaries contain aldose reductase, whereas the capillary endothelial cells had no detectable levels of the enzyme[8]. Although retinal

The Effects of Aging and Environment on Vision
Edited by D.A. Armstrong *et al.*, Plenum Press, New York, 1991

changes in animal models of diabetes do not strictly parallel those in human diabetic retinopathy[9], several aldose reductase inhibitors have been shown to inhibit the retinal capillary basement membrane thickening that develops in both diabetic and galactosemic rats[5,6]. The specific biochemical link(s) between diabetic retinopathy, aldose reductase activity, and basement membrane thickening have not yet been elucidated. However, the evidence for such a link was recently strengthened by the report that one aldose reductase inhibitor can significantly improve retinal function in diabetic rats, as assessed by the c-wave of the electroretinogram[10].

Cataract and retinopathy are the two major, but not the only, manifestations of ocular tissue dysfunction in human diabetics; increased risks of glaucoma and impaired corneal wound-healing, observed following vitrectomy, have been documented[1]. Although the study of these complications in animal models of diabetes is difficult, there is evidence that aldose reductase inhibitors improve corneal wound-healing in diabetics[11,12]. Also we have recently demonstrated sorbitol accumulation in human trabecular meshwork cells incubated in high glucose medium and the inhibition of this response by known aldose reductase inhibitors[13]. These data suggest that aldose reductase inhibitors may be useful therapeutic agents for glaucoma in human diabetics.

Many of the functional and morphological changes of diabetes in humans and the animal models are characteristic of an accelerated aging process which inhibits growth and compromises tissue function and integrity. For example, cataract in humans has a higher incidence among the elderly (hence, the designation "senile" cataract), and is often associated with a decreased growth rate of the lens[14]. Growth of lenses of severely diabetic rats is similarly inhibited, concomitant with progression of cataract[15]. Increased flux of glucose through the sorbitol pathway in the lens epithelial cells can account for this effect on growth: ATP is diverted from biosynthetic pathways into ATP-driven Na^+ and K^+ exchange and other "pump" activities, as a mechanism to lower the osmolality within the cell.

Vracko has published evidence that thickening of basement membrane, in particular the homogeneous basal lamina component, is an age-dependent phenomenon that occurs only in certain anatomic sites and is accelerated by diabetes[16]. Data from controlled animal studies and from human autopsy samples supported the hypothesis that cells of certain tissues, such as muscle capillaries and renal tubules, deposit a new layer of basal lamina in response to injury[16]. Vracko has postulated that the deposition of multiple layers of basal lamina results from an increased rate of cell death, followed by movement of new cells into the residual extracellular matrix. The new cells synthesize additional basal lamina which may fuse smoothly with the existing matrix, or may form discontinous clefts due to trapped cell debris. The morphological features of these deposits of basal lamina, which contribute to basement membrane thickening, appear identical in corresponding tissues of humans and animal models[16].

The present study was undertaken to define the effects of a new aldose reductase inhibitor AL-1567 (racemic) and its two resolved isomers on some biochemical, functional, and ultrastructural alterations in the lens and retina of severely diabetic rats. To the extent that the diabetic rat model is a highly exaggerated model of aging, in which normal regulatory controls over metabolism are severely impaired, and which develops aging-like characteristics in various tissues, the results suggest that it may be possible to identify specific biochemical processes linked to aging and to design therapeutic agents that can prevent or intervene in the pathophysiological changes associated with aging.

EXPERIMENTAL

Aldose reductase activity was determined by a standard spectrophotometric assay[17,18], with a soluble rat lens enzyme preparation. The inhibitor concentration inhibiting the rate by 50% (IC_{50} value) was computed from linear regression analysis of the linear portion of a plot of (% inhibition) vs. log inhibitor concentration. For assay of aldose reductase and sorbitol dehydrogenase activities in lenses and retinas of individual rats, sensitive fluorescence assays which have been described in detail[18] were employed.

Male Sprague/Dawley rats (SASCO, Omaha, NE), weighing 230-300 g, were made diabetic by intravenous injection of 50 mg/kg streptozotocin (Sigma Chemical Co), freshly dissolved in 0.05 Molar citric acid buffer, pH 4.5, containing 0.5% NaCl. One week later, blood was collected from all animals by tail venupuncture into heparinized capillary tubes; plasma glucose was assayed by a standard coupled enzymatic assay (glucose oxidase/peroxidase method) obtained from Sigma. Injected rats with non-fasting plasma glucose values greater than 400 mg/dL were selected for the study, and randomly assigned to study groups. Throughout the study, rats had unrestricted access to water and Purina Rodent Chow.

In the acute model, daily oral dosing with the aldose reductase inhibitors was initiated 2.5 to 3 weeks after induction of diabetes, when tissue sorbitol levels have attained their maximal values[15]. After 8 days of treatment, the animals were sacrificed and the tissues frozen immediately in a dry ice/acetone bath for subsequent sorbitol analyses. For assay of enzyme activities, a high-speed supernatant was prepared from tissues homogenized in 3 mL/g wet weight of 0.1 Molar potassium phosphate buffer, pH 6.8, containing 5 mMolar mercaptoethanol. The eyes were trimmed of extraneous tissue and carefully dissected by an anterior approach, i.e., by an incision almost completely around the limbus. After removal of the lens and vitreous, the remaining tissue, designated as retina, was rinsed with saline, carefully blotted, and weighed.

Sorbitol was determined by a fluorometric procedure[19] in protein-free tissue extracts prepared by standard procedures[20], with $ZnSO_4$ and $Ba(OH)_2$. The reaction mixtures contained 1.0 mL of 0.05 Molar glycine buffer, pH 9.4, 2 U sorbitol dehydrogenase, 0.8 mg NAD^+, and 0.5 mL of the original or diluted tissue extract; sorbitol dehydrogenase was omitted from sample blanks. The increase in fluorescence after a 20 minute incubation was measured with a Perkin-Elmer 650-10S fluorometer, set for 365 nm excitation and 455 nm emission wavelengths, and compared with identically treated sorbitol standards. Sorbitol dehydrogenase and NAD^+ were supplied by Sigma.

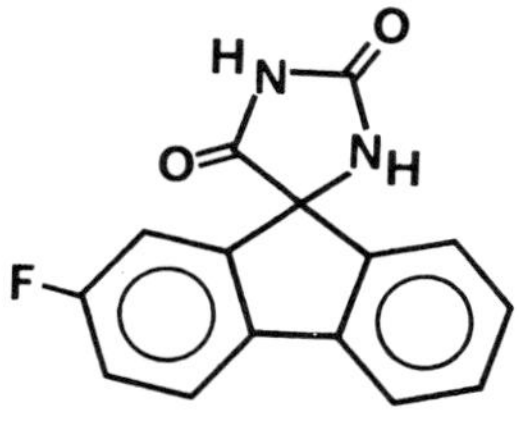

Figure 1. Chemical structure of AL-1567 (racemic mixture).

Table 1. Mean Sorbitol Values in Lenses of 3-Week Diabetic Rats Dosed for 8 days with AL-1567 (racemic) or its Isomers.

Status	Compound	Dose, mg/kg/day	Sorbitol,[a] nmol/mg wet weight	Sorbitol, % diabetic control
Diabetic	-------	-----	40.3 ± 4.79	100.
Diabetic	(-) AL-1567	0.5	23.3 ± 6.30	57.8
Diabetic	(±) AL-1567	0.5	25.0 ± 4.70	62.0
Diabetic	(+) AL-1567	0.5	41.2 ± 2.76	102.
Diabetic	(-) AL-1567	1.5	3.13 ± 1.73	7.8
Diabetic	(±) AL-1567	1.5	6.90 ± 2.00	17.1
Diabetic	(+) AL-1567	1.5	22.6 ± 4.60	56.1
Diabetic	(±) AL-1567	4.0	2.06 ± 0.39	5.1
Normal	-------	-----	0.34 ± 0.05	0.8

[a] Mean value ± S.D., N = 6.

The chronic ocular effects of AL-1567 were assessed in diabetic rats dosed orally with various levels of the compound for a period of 14 weeks, beginning one week after induction of diabetes. The cataract score was assigned weekly by an observer without knowledge of the identification or prior score of each animal; the score approximated the area of observable lens changes, with a score of 6 corresponding to full nuclear opacity. Measurement of the thickness of retinal capillary basement membranes was performed on samples from a randomly-selected subset of rats in each group, prepared and analyzed by procedures described elsewhere[21].

RESULTS

AL-1567 (Fig. 1) and its isomers could be distinguished as inhibitors of aldose reductase, both in vitro and in vivo: IC_{50} values were 2.7, 3.6, and 6.7 x 10^{-8} Molar for (-)AL-1567, AL-1567, and (+)AL-1567, respectively. The racemic mixture had intermediate activity, as expected; the (-) isomer was about 2.5 times more active than the (+) isomer. The lens activities of these compounds in vivo, evaluated by inhibition of lens sorbitol accumulation in acutely-dosed diabetic rats, are compared in Table 1. The (+) isomer was clearly least active; the activity difference between the racemic mixture and the (-) isomer showed the expected trend, but was not statistically significant. Since 1.5 mg/kg of (+) AL-1567 produced inhibition of lens sorbitol equivalent to that of the 0.5 mg/kg dose of (-) AL-1567, the lens activity ratio of the isomers is approximately 3, which agrees well with the in vitro activity ratio of 2.5.

In this same study, all three compounds were also able to inhibit sorbitol accumulation in retinas of diabetic rats, Fig. 2. It should be noted that the absolute amounts of sorbitol in both normal and untreated diabetic rat retinas are measurably less than in lenses of the respective groups. However, the mean retinal sorbitol level of the untreated diabetic rats is elevated significantly with respect to the normal value. At a dose level of 1.5 mg/kg, the racemic mixture and both isomers of AL-1567 produced considerable inhibition of retinal sorbitol; moreover, there was a qualitative correlation of the activities of the three compounds in diabetic rat retina with their lens activity data, with (+) AL-1567 displaying less activity than the (-) isomer or the racemic compound.

The effects of various doses of AL-1567 (racemic) on cataract progression in chronically diabetic rats are shown in Fig. 3. The first lens changes associated with cataract were observed in the diabetic control group 5 to 6 weeks after induction of diabetes; after 15 weeks, the mean cataract score of this group increased to a value greater than 5, corresponding to almost complete nuclear opacity. Daily, oral dosing of AL-1567, initiated one week after streptozotocin injection, completely prevented any appearance of cataractous changes at doses greater than or equal to 1.5 mg/kg. Moreover, the lowest dose tested, 0.5 mg/kg, had a significant inhibitory effect on cataract: the final (15-week) mean cataract score of 0.5 for this group corresponds to the minimal observable change at the lens equator. From the data in Table 1, it can be seen that the 0.5 mg/kg dose of AL-1567 (racemic) approximates the ED_{50} value of this compound for inhibition of lens sorbitol in the acutely-dosed diabetic rat model.

Table 2 summarizes the data on retinal capillary basement membrane thickness determined for the various groups in the AL-1567 chronic dosing study. Details of these procedures will be published elsewhere. It should be noted that all values reported represent an average of 10 separate capillaries for each animal (random selection), and the micrographs were analyzed in a masked fashion. Two distinct layers of basement membrane could be discerned, an "inner" layer surrounding the endothelial cells, and an "outer" layer; in most regions, these two layers were clearly separated by cell mass, originating most likely from pericytes. Measurements of thickness of the capillary basement membranes were made only where the two membranes were not contiguous. As shown in Table 2, the outer membrane was thicker in normals, displayed both a larger absolute and percentage increase in 15-week diabetic rats, and appeared to be somewhat less responsive to AL-1567 than

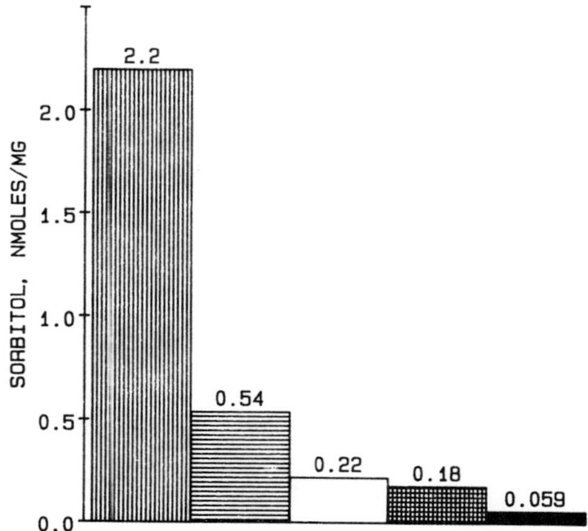

Fig. 2. Inhibition of retinal sorbitol accumulation in 3-week diabetic rats dosed for 8 days with AL-1567 (racemic) or its isomers. The diabetic and normal control values are the extreme left and right bars, respectively. Treated diabetic rats received a dose of 1.5 mg/kg/day of (+) AL-1567 ( ), (±) AL-1567 ( ) or (-) AL-1567 ( ). Values above the bars give the actual sorbitol concentrations.

Table 2. Mean Values of Thickness of Retinal Capillary Inner and Outer Basement Membranes of Diabetic Rats Dosed 14 Weeks with AL-1567.

Status	Dose, mg/kg/day	N	Inner Membrane[a]	Outer Membrane[a]
Diabetic	-----	5	821 ± 37	1125 ± 41
Diabetic	0.5	5	720 ± 43	1064 ± 48
Diabetic	1.5	5	710 ± 16	1015 ± 32
Diabetic	4.0	3	718 ± 19	945 ± 3
Diabetic	16.0	5	708 ± 2	1003 ± 35
Normal	-----	7	698 ± 23	915 ± 30

[a] In Angstroms, ± S.E.M.

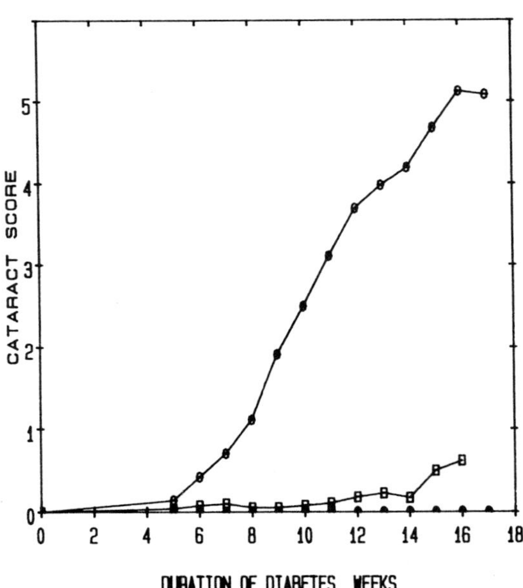

Fig. 3. Inhibition of cataract in chronically diabetic rats by various doses of (±) AL-1567. The cataract scores ranged from 0 (normal) to 6 (nuclear opacity). Diabetic control (untreated) group, (O); normal group, (●); 0.5 mg/kg/day dose level, (□); 1.5, 4.0, and 16.0 mg/kg/day dose levels, (●), identical to normal group.

Table 3. Comparison of Aldose Reductase and Sorbitol Dehydrogenase Specific Activities and Activity Ratios in Rat Lens and Retina.

Tissue	Aldose reductase (AR) specific activity[a]	Sorbitol dehydrogenase (SDH) specific activity[a]	AR/SDH
Lens	0.55	0.12	4.62
Retina	0.96	0.99	0.97

[a] Units, nmol x min $^{-1}$ x (mg protein) $^{-1}$.

the inner membrane. However, a dose level of 4.0 mg/kg AL-1567 normalized the thickness of both basement membrane layers of retinal capillaries of severely diabetic rats.

The data in Table 1 and Fig. 2 indicate that the lens and retinal sorbitol concentrations of the diabetic control group are quite different, approximately 40 and 2.2 nmol/mg tissue, respectively. Sorbitol accumulation under hyperglycemic conditions requires that the rate of sorbitol formation via aldose reductase catalysis exceed the rate of metabolism, mediated by sorbitol dehydrogenase[1] (cf. Eq.(1)). Thus, the ratio of aldose reductase to sorbitol dehydrogenase activity in a given tissue, reflecting the balance between production and disappearance of sorbitol, should be a good indicator of the polyol-linked osmotic stress potential under hyperglycemic conditions in that tissue. The aldose reductase and sorbitol dehydrogenase activities in rat lens and retina were measured, with the results given in Table 3. Although the specific activity of aldose reductase in retina is somewhat larger, the ratio of aldose reductase to sorbitol dehydrogenase activity is larger in lens than retina by a factor of about 5. Consistent with this result, the ratio of sorbitol levels in lens and retina of diabetic rats is even larger, approximately 20 (cf. Table 1 and Fig.2). It should be noted that the steady-state value of sorbitol attained in a given tissue will also depend on the duration and severity of glucose elevation and the availability of pyridine nucleotides in the proper oxidation state to support both aldose reductase and sorbitol dehydrogenase reactions (cf. Eq.(1)). However, these results suggest that the ratio of aldose reductase to sorbitol dehydrogenase in each tissue is an important determinant of the peak sorbitol levels, and, thus, an indicator of the extent of the osmotic and metabolic stress that will develop in that tissue during hyperglycemia.

The specific activities of both enzymes in lens appear to be quite low; however, in lens, aldose reductase appears to be localized in cells of the epithelial and outer cortical layers[22,23]. Thus, the lens cells which are metabolically most active experience the brunt of the polyol-mediated osmotic stress under hyperglycemic conditions. Similarly, aldose reductase in human retinal capillaries has been reported to be localized in pericytes[8], the cells which are selectively lost in human diabetic retinopathy[7]. Thus, tissue average sorbitol values or aldose reductase/sorbitol dehydrogenase ratios, such as those reported in this study, may underestimate the severity of the osmotic and metabolic stress experienced by certain susceptible cell types during hyperglycemia.

DISCUSSION

The results of these studies provide additional support for the involvement of aldose reductase in early metabolic changes that evolve into the end-stage pathology associated with ocular complications of diabetic rats[1-3]. Daily oral administration of low doses of AL-1567 for a period of eight days decreased in a dose-response manner lens and retinal sorbitol levels which were significantly elevated above normal values by three prior weeks of severe hyperglycemia. It was shown that the relative in vivo and in vitro aldose reductase activities of (racemic) AL-1567 and its isomers correlated well. Dose levels of (racemic) AL-1567 which inhibited lens sorbitol accumulation in the acute study also effectively inhibited progression of gross cataract morphology in diabetic rats treated with the compound daily for a period of 14 weeks. Since AL-1567 does not alter plasma or tissue glucose levels[24], these data corroborate other studies which implicate sorbitol and not glucose as the cataractogenic agent in severely diabetic rats[1-3]. Other documented biochemical and morphological features of diabetic cataract appear to develop after the early increase in lens sorbitol concentration[1,2]. The activities of several other aldose reductase inhibitors in similar acutely-dosed diabetic rat models have been reported [3,25-28]; of these, only Sorbinil appears to have measurable activity in lens, with an ED_{50} for inhibition of lens sorbitol of approximately 1.5 mg/kg/day[28]. Our data indicate tht (R)(-) AL-1567 is about three times more active as an anti-cataract agent than either its isomer or Sorbinil, which has (S)(+) configuration.

Although capillary basement membrane appears to be thicker than normal in all postpubertal human diabetics[29], the factors which contribute to this phenomenon and its pathological significance for any diabetic vascular complication, such as retinopathy or nephropathy, are not understood. However, a link between aldose reductase activity and retinal capillary basement membrane thickening in animal models of diabetes has been established by our studies and by others[6,30] with different aldose reductase inhibitors. In an effort to identify this link, Williamson and colleagues have characterized extensively early alterations in vascular tissue barrir integrity, i.e., increased permeation of ^{125}I-albumin, in various tissues of diabetic rats[31-33]. Their work implicates a compromised barrier function of the capillary vasculature as an important functional link between the early metabolic alteration (sorbitol accumulation) and the late-developing morphological change (basement membrane thickening) of diabetic rats[31-33]. These changes in vascular function, as well as elevated tissue sorbitol levels, of the diabetic animals were normalized by several aldose reductase inhibitors and were also significantly decreased by castration of one group of animals several weeks prior to streptozotocin injection[33]. The potentiation of diabetes-induced vascular tissue dysfunction in rats by circulating levels of androgens observed in these studies[33] is supported by evidence that puberty exacerbates diabetic complications in males [34,35]. One mechanism by which androgens might exert this effect is suggested by recent reports that puberty produces a 2- to 3-fold increase in aldose reductase activity in lens, sciatic nerve[15a] and kidney[36] of male rats. The available data suggest that basement membrane thickening is a response to, rather than a cause of, the vascular pathophysiology of diabetes, in support of Vracko's hypothesis[16] that injury to vascular tissue provides the stimulus for excessive production of basement membrane. These results appear to implicate aldose reductase activity in the early phase of vascular injury in diabetic animals, but additional studies are required to elucidate the specific biochemical mechanism of the increase in vascular permeation of albumin.

The use of aldose reductase inhibitors in animal models of diabetes to probe early, functional consequences of increased sorbitol pathway activity should increase our understanding of the molecular basis of the pathophysiology of diabetic complications in ocular tissues. Since these complications resemble those associated with aging, we expect that these studies will reveal valuable information about the effects of aging on ocular tissues and will stimulate research into other animal models of aging and other rational therapeutic approaches to preserving visual function in the elderly.

REFERENCES

1. P.F. Kador, W.G. Robinson, Jr., and J.H. Kinoshita, 1979, The pharmacology of aldose reductase inhibitors, in: "Annual Review of Pharmacology and Toxicology," R. George, R. Okun, and A.K. Cho, ed., Annual Reviews, Inc., Palo Alto, CA.

2. J.H. Kinoshita, S. Fukushi, P. Kador, and L.O. Merola, 1979, Aldose reductase in diabetic complications of the eye, Metabolism 28:462.

3. D. Dvornik, 1987, Aldose reductase inhibition. An approach to prevention of diabetic complications, D. Porte, Ed. New York, Biomedical Information Corp. (McGraw-Hill).

4. B.W. Griffin, M.L. Chandler, and L. DeSantis, 1984, Prevention of diabetic cataract and neuropathy in rats by two new aldose reductase inhibitors, Invest. Ophthalmol. Vis.Sci. 25 (Suppl):136.

5. M.L. Chandler, W.A. Shannon, and L. DeSantis, 1984, Prevention of retinal capillary basement membrane thickening in diabetic rats by aldose reductase inhibitors, Invest. Ophthalmol. Vis. Sci. 25 (Suppl):159.

6. W.G. Robison, Jr., P.F. Kador, Y. Akagi, J.H. Kinoshita, R. Gonzalez, and D. Dvornik, 1986, Prevention of basement membrane thickening in retinal capillaries by a novel inhibitor of aldose reductase, Tolrestat, Diabetes 35:295 (1986).

7. J.R. Williamson and C. Kilo, 1976, Basement-membrane thickening and diabetic microangiopathy, Diabetes 25:925.

8. Y. Akagi, P.F. Kador, T. Kuwabara, and J.H. Kinoshita, Aldose reductase localization in human retinal mural cells, Invest. Ophthalmol. Vis. Sci. 24:1516.

9. R.G. Tilton, L.S. LaRose, C. Kilo, and J.R. Williamson, 1986, Absence of degenerative changes in retinal and uveal capillary pericytes in diabetic rats, Invest. Ophthalmol. Vis. Sci. 27:716.

10. L.C. MacGregor and F.M. Matschinsky, Treatment with aldose reductase inhibitor or with myo-inositol arrests deterioration of the electroretinogram of diabetic rats, J. Clin. Invest. 76:887.

11. S. Fukushi, L.O. Merola, M. Tanaka, M. Datiles, and J.H. Kinoshita, 1980, Reepithelialization of denuded corneas in diabetic rats, Exp. Eye Res. 31:611.

12. L.A. Meyer, K.M. Foley, J.L. Ubels, and H.F. Edelhauser, 1986, Diabetes-induced corneal endothelial changes in rats, Invest.Ophthalmol. Vis. Sci. 27 (Suppl):175.

13. B.M. York, B.W. Griffin, J.A. Alvarado, R. Kurtz, and J.R. Polansky, 1986, Sorbitol accumulation by human trabecular meshwork cells cultured in high glucose medium and inhibition by potent aldose reductase inhibitors, 7th International Congress of Eye Research (Abstract), Nagoya.

14. O. Hockwin, U. Eckerskorn, W. Schmidtmann, V. Dragomirescu, I. Korte, and H. Laser, 1984, Epidemiological study of the association between lens cataract and case history, blood composition, and enzymes involved in lens carbohydrate metabolism, Lens Research 2:23.

15. H. Kuriyama, K. Sasaki, and M. Fukuda, 1983, Studies on diabetic cataract in rats induced by streptozotocin II. Biochemical examinations of rat lenses in relation to cataract stages, Ophthalmic Res. 15:191.

15a. B.W. Griffin and L.G. McNatt, 1987, Measurement of the sorbitol pathway enzymes in tissues of normal and galactosemic rats, Invest. Ophthalmol. Vis. Sci. 28 (Suppl): 78.

16. R. Vracko, R.E. Pecoraro, and W.B. Carter, 1980, Thickness of basal lamina of epidermis, muscle fibers, muscle capillaries, and renal tubules: changes with aging and in diabetes mellitus, Ultrastruct. Path. 1:559.

17. S. Hayman and J.H. Kinoshita, 1965, Isolation and proeprties of lens aldose reductase, J. Biol. Chem. 240:877.

18. B.W. Griffin and L. M. McNatt, 1986, Characterization of the reduction of 3-acetylpyridine adenine dinucleotide phosphate by benzyl alcohol catalyzed by aldose reductase, Arch. Biochem. Biophys. 246:75.

19. J.I. Malone, G. Knox, S. Benford, and T.A. Tedesco, 1980, Red cell sorbitol. An indicator of diabetic control. Diabetes 29:861.

20. W.A. Sherman and M.A. Stewart, 1966, Identification of sorbitol in mammalian nerve, Biochem. Biophys. Res. Commun. 22:492.

21. W.A. Shannon, Jr., D.L. Rockholt, and S.B. Bates, 1982, Computer-assisted measurement of the thickness of biological structures, Comput. Biol. Med. 12:149.

22. Y. Akagi, Y. Yajima, P.F. Kador, T. Kuwabara, and J.H. Kinoshita, 1984, Localization of aldose reductase in the human eye, Diabetes 33:563.

23. J.A. Jedziniak, L.T. Chylack, Jr., H.-M. Cheng, M.K. Gillis, A.A. Kalustian, and W.H. Tung, 1981, The sorbitol pathway in the human lens: aldose reductase and polyol dehydrogenase, Invest. Ophthalmol. Vis. Sci. 20:314.

24. B.W. Griffin, L.G. McNatt, M.L. Chandler, and B.M. York, 1987, Effects of two new aldose reductase inhibitors, AL-1567 and AL-1576, in diabetic rats, Metabolism 36:486.

25. R. Kikkama, I. Hatamaka, H. Yasuda, N. Kobayashi, Y. Shigeta, H. Terashima, T. Morimura, and M. Tsuboshima, 1983, Effect of a new aldose reductase inhibitor, (E)-3-carboxymethyl-5-[(2E)-methyl-3-phenylpropenylidene] rhodanine (ONO-2235) on peripheral nerve disorders in streptozotocin-diabetic rats, Diabetologia 24:290.

26. N. Simard-Duquesne, E. Greselin, J. Dubuc, and D. Dvornik, 1985, The effects of a new aldose reductase inhibitor (Tolrestat) in galactosemic and diabetic rats, Metabolism 34:995.

27. D. Stribling, D.J. Mirrlees, H.E. Harrison, and D.C.N. Earl, 1985, Properties of ICI 128,436, a novel aldose reductase inhibitor, and its effects on diabetic complications in the rat. Metabolism 34:336.

28. J.I. Malone, H. Leavengood, M.J. Peterson, M.M. O'Brien, M.G. Page, and C.E. Aldinger, 1984, Red blood cell sorbitol as an indicator of polyol pathway activity. Inhibition by Sorbinil in insulin-dependent diabetic subjects, Diabetes 33:45.

29. R.G. Tilton, P.F. Hoffman, C. Kilo, and J.R. Williamson, 1981, Pericyte degeneration and basement membrane thickening in skeletal muscle capillaries of human diabetics, Diabetes 30:326.

30. W.G. Robison, Jr., P.F. Kador, and J.H. Kinoshita, 1983, Retinal capillaries: basement membrane thickening by galactosemia prevented with aldose reductase inhibitor, Science 221:1177.

31. J.R. Williamson, K. Chang, E. Rowold, J. Marvel, M. Tomlinson, W.R. Sherman, K.E. Ackermann, and C. Kilo, 1985, Sorbinil prevents diabetes-induced increases in vascular permeability but does not alter collagen cross-linking, Diabetes 34:703.

32. J.R. Williamson, K. Chang, E. Rowold, J. Marvel, M. Tomlinson, W.R. Sherman, K.E. Ackerman, and C. Kilo, 1986, Diabetes-induced increases in vascular permeability and changes in granulation tissue levels of sorbitol, myo-inositol, chiro-inositol, and scyllo-inositol are prevented by Sorbinil, Metabolism 35 (Suppl 1):41.

33. J.R. Williamson, K. Chang, R.G. Tilton, C. Prater, J.R. Jeffrey, C. Weigel, W.R. Sherman, D.M. Eades, and C. Kilo, 1987, Increased vascular permeability in spontaneously diabetic BB/W rats and in rats with mild versus severe streptozotocin-induced diabetes. Prevention by aldose reductase inhibitors and castration, Diabetes 36:813.

34. F.I. Caird, A. Ferie, T.G. Ramwell, 1969, The natural history of diabetic retinopathy, in Diabetes and the Eye, Oxford, UK, Blackwell. pp. 72-100.

35. R.N. Frank, 1984, On the pathogenesis of diabetic retinopathy, Ophthalmology 91:626.

36. M. Sochar, S. Kunjara, A.L. Greenbaum, and P. McLean, Renal hypertrophy in experimental diabetes. Effect of diabetes on the pathways of glucose metabolism: differential response in adult and immature rats, Biochem. J. 234:573.

QUESTIONING THE NATURE OF AGE PIGMENT (LIPOFUSCIN) IN THE HUMAN RETINAL PIGMENT EPITHELIUM AND ITS RELATIONSHIP TO AGE-RELATED MACULAR DEGENERATION

Graig Eldred

University of Missouri-Columbia
School of Medicine
Department of Ophthalmology
Columbia, Missouri 65212

INTRODUCTION

Second only to cataracts, age-related macular degeneration (AMD) is a leading cause of visual impairment in the elderly. Incidence of AMD has been estimated at between 1.3% and 4% of the population in the 65-74 year age groups (18, 29). With both the number and proportion of people over the age of 65 continuing to increase (27), AMD is likely to become an increasingly significant health problem. There is reason to suspect that age-related changes in the retina may be at the root of this disease (45,46,10). In the following discussion, the reasoning, the resulting theory, a discrepancy in that theory, and current experimental progress into the resolution of this discrepancy will be described.

CLINICAL MANIFESTATIONS AND TREATMENT OF AMD

Clinically, AMD manifests itself as a loss of vision in the central visual field. In advanced hemorrhagic or disciform AMD, for reasons that are unknown, blood vessels break through a thick matrix barrier (Bruch's membrane) and begin to grow into the retina and vitreous humor from the underlying vascular choroidal layer. With time these vessels leak, become fibrotic and scar tissues begin to pull the neural retina out of position leading to retinal cell death (34).

There exists no known prevention or cure for this disease. Once detected, early progression can be stemmed through laser therapy (30). In this treatment a laser beam is targeted at the melanin pigment granules of a monolayer of neuroepithelial cells underlying the retina, the retinal pigment epithelium (RPE). These melanin granules absorb the laser's photic energy and rapidly convert it into heat manifested as a pinpoint thermomechanical eruption which destroys the surrounding tissues. By treating in the direct vicinity of the expanding blood vessels, the new vessel growth and hemorrhage may be halted thereby delaying the progress of the disease.

The Effects of Aging and Environment on Vision
Edited by D.A. Armstrong *et al.*, Plenum Press, New York, 1991

THEORY OF AN AMD-LIPOFUSCIN RELATIONSHIP

What causes these vessels to begin to break through Bruch's membrane is uncertain. The membrane itself undergoes many age-related changes which are only now beginning to be well characterized (16, 36). These changes, in turn, seem to be preceded by a major ultrastructurally visible change in the overlying RPE. The RPE cells become heavily laden with age pigments (lipofuscin granules) (17).

RPE cells are laid down at the time of birth and reproduce rarely, if ever (48). Over time they accumulate large number of lipiodal granules in their cytoplasm. At the same time, the number of melanin granules declines (17). The melanin and age pigment are quite distinct and it is doubtful that the melanin granules are simply transformed into lipofuscin (20, 21).

It has been theorized that as the RPE cells become constipated with lipofuscin granules, they attempt to rid themselves of excess cell volume through the mechanism of apoptosis (3). In this theory, the cells pinch off a portion of their cytoplasm basally where it remains deposited in the inert Bruch's membrane. The debris eventually mounds up to form drusen, and the RPE-derived materials may serve as nucleation sites for calcification. It may be in these areas of deteriorating debris that choroidal-RPE-retinal nutrient exchange is impaired, an angiogenesis factor is secreted from a yet to be identified source, and neovascularization begins.

APPROACHES TO THEORY VALIDATION

The causal connections described above between RPE lipofuscin granule accumulation, age-related changes in Bruch's membrane, and age-related macular degeneration (AMD) are deduced through morphological evidence but are not yet directly approachable through experimental means. No experimental animal model for this disease exists. No approach for manipulating the levels of age pigment within the RPE exists. To be able to correlate the incidence and severity of AMD with the levels of lipofuscin in the RPE, a means of manipulating the granule burden independent of any other experimental parameters is needed. Vitamin E deficiency leads to an accumulation of pigment granules in the RPE which appear very similar to the natural age pigment (25), but their compositions may not be the same and other retinal or vascular parameters may be altered in the E-deficient state.

To be able to manipulate lipofuscin granule numbers, the molecular composition of these pigments must be known. Currently, the nature of the granule contents may only be deduced through in vitro evidence. Compositional analysis is the natural starting place for clarification, and this is the goal of current investigations. Knowledge of the specific molecular composition of the granules would provide a basis upon which to design experiments aimed at removing or manipulating the levels of specific compounds and thereby the umber of lipofuscin granules present. From here any role for lipofuscin granules in human AMD would be clarified.

THEORY OF LIPOFUSCINOGENESIS IN THE RPE

To begin a discussion of the deductive logic as to what the accumulating molecules might be, an understanding of the biology of the RPE is required. The RPE is located between the neural retina and the vascular choroidal layer. It has many functions in nutrient and metabolite exchange between the photoceptor cells of the retina and the vascular supply in the choroid (47). The melanin granules of the RPE also play a role in the absorption of stray light which would otherwise fog the visual image.

The function most pertinent to the story of lipofuscin generation deals with a phenomenon occurring in the overlying photoceptor cells. The pigment(s) responsible for the primary photochemical events in visual transduction lie embedded in stacked membranous disks of the photoreceptor outer segment. These disks are continually renewed at the inner end of the stack (farthest from the RPE cell). At the distal tip of the stack which lies conveniently within the waiting arms of the RPE cell, disks are shed in a circadian cycle (2). As packets of these disks break off the RPE cell engulfs them via phagocytosis and subjects them to this intracellular digestive processes. The lipid, protein, and vitamin A building blocks are freed, metabolized, and laminated or recycled.

It is at this point that the system is believed to break down. It is hypothesized that some of the photoreceptor outer segment disk membrane lipids may become oxidized prior to engulfment (15). Some of the breakdown products, in particular, malondialdehyde, a three carbon dialdehyde tautomer, are capable of crosslinking any amine containing compound. This gluing together or chemical fixation of materials would render them resilient to digestion by the lysosomal enzymes. Hence, undigestible materials would gradually start to accumulate within the lysosomal system and appear as tertiary lysosomes (i.e., residual bodies, lipofuscin granules, age pigments). That these materials are within the lysosomal compartment and are resilient to enzymatic attack is evidenced by positive histochemical staining for lysosomal enzymes (14).

What would cause these lipids to become oxidized? Three main factors would make the photoreceptor outer segments particularly susceptible to autoxidative processes---their fatty acid composition, the presence of abundant light and oxygen, and the presence of a possible photosensitizer (22). The photoreceptor outer segment lipids are unusual in that they contain a large amount of polyunsaturated fatty acids of a kind which possess six unsaturated double bonds (docosahexaenoic acid, 22:6w3) (1). Polyunsaturated fatty acids in general are particularly susceptible to free radical medicated autoxidative degration (42), and docosahexaenoic acid is the most highly unsaturated acyl group found in membrane phospholipids of mammalian tissues (26). To start the degradative process, an initiation event must occur. The visual pigments themselves may act as photosensitizing agents (6). When a rhodopsin molecule absorbs a photon, it briefly exists in an excited state. Under certain conditions, and in the presence of molecular oxygen, an electron may be transferred from the excited pigment molecule to oxygen. This activated oxygen may undergo a variety of reactions resulting in highly reactive free radical and activated species which can in turn initiate the free radical lipid autoxidation reactions. Thus, the retina has all ingredients necessary for this scenario to come to pass. That this is a rather rare event is assured by the presence of a variety of free radical chain reaction terminators such as vitamin E and superoxide dismutase (23). Were it a very common event, one would expect to see lipofuscin accumulation at a much earlier age.

DISCREPANCY IN THE THEORY OF LIPOFUSCINOGENESIS

The evidence cited makes this a very appealing mechanism and working hypothesis for the formation of lipofuscin. There is one very disturbing inconsistency in this theory, however, which deals with fluorescence. Lipofuscin granules from all tissues emit a golden-yellow to orange fluorescence (31). Indeed, this auto fluorescence is pathognomonic for lipofuscin. The materials that accumulate within the granules must account for this fluorescence.

By-products of lipid autoxidation, both in the presence and absence of amine-containing compounds, are known to fluoresce (8, 38). In 1975, it was reported that the fluorescence excitation and emission spectra of both human age pigment extracts and in vitro lipid autoxidation byproducts are essentially identical (41). From there it was concluded that the fluorescent molecules were categorically the same. The molecular structure responsible for the fluorescence was purported to be an imine-conjugated Schiff base (ICSB) linkage (-N=CH-CH=CH-NH-), and compounds synthesized around this structure were established as standards for use in assays for both lipid autoxidation and age pigment accumulation (32, 33). Indeed, the term lipofuscin has come to be used interchangeably by some investigators for both age pigment and fluorescent lipid autoxidation byproducts.

However, in 1982, it was demonstrated that the similarities between their spectra no longer exist if the instrumentation is properly calibrated for its color blindness (i.e., spectral sensitivity) (13). Spectrally corrected total luminescence spectra of the chloroform soluble fluorophores of aged human RPE do not compare with similar spectra of diglycinyl ICSB (10). This latter is one of the compounds synthesized, spectrally characterized and recommended for use as a standard in lipofuscin and lipid autoxidation research by Tappel and collaborators (4).

EXPLANATIONS FOR THE ORIGIN OF THE DISCREPANCY

It is fair to ask now could such misinterpretations have arisen in the first place.

One explanation offered for the difference between the yellow emissions seen by microscopists and the blue emissions seen by spectroscopists deals with spectral shifts caused by environmental or solvent effects. Changes in solvent polarity, etc. are known to be capable of causing changes in both absorbance and emission peaks. These shifts are well known for a variety of solvents and are dependent upon the molecular structure of the compound (40). The shifts are never more than ten nm, however, and cannot account for the difference of 150 nm between blue wavelengths, centered around 420 nm, and yellow wavelengths, narrowly centered around 575 nm.

Internal quenching is another phenomenon which could have explained an apparent discrepancy between the yellow emissions of the lipofuscin granules seen in fluorescence microscope and the blue fluorescent emission measured by investigators in fluorescence spectrometry of chloroform extracts. This also has an important bearing on proper procedures for measuring the fluorescence spectra of any molecule extracted from lipofuscin. The excitation and emission peaks of many fluorescent compounds are not widely separated. In these cases the excitation curve and the emission curves may overlap, especially when highly concentrated. When this occurs, the low wavelength emissions may be self-absorbed by neighboring molecules before having a chance to exit the granule or cuvette, whichever the case may be. This results in a measured spectrum or observed emission with a peak emission at artifactually high wavelengths; the low wavelength emissions are masked. For this reason, proper fluorescence spectroscopy protocols mandate that all solutions be diluted to less than 0.1 absorbance unit prior to analysis.

Could it be that blue-emitting lipid autoxidation products are concentrated within the lipofuscin granules, and that internal quenching phenomena allow only the yellow emissions apparent by fluorescence microscopy? In our work with the lipofuscin of RPE, working with both crude extracts and fluorophores separated by thin-layer chromatography, this has not proven to be the case. The fractions thus far isolated all emit at wavelengths far outside the blue emission range when the possibility of internal quenching is eliminated. In working with whole RPE cell extracts from individual donors, blue emitters are seen only occasionally and are now interpreted to be drug metabolites.

Problems with biased instrumental spectral sensitivity have already been mentioned, and herein lies the most likely explanation for the origin of the discrepancy which has evolved in this field. Using spectrofluorometric instrumentation that has not been spectrally corrected, a blue fluorescence emission peak can be measured from chloroform soluble extracts of aged, lipofuscin-laden human RPE. Yet when viewed under the same light outside the instrument, the sample emits a yellow fluorescence.

To correct for this bias several procedures are available, one of which involves the use of special fluorescent compounds whose quantum efficiency is independent of wavelength over an extended range (i.e., a quantum counter) (37). While the procedural details are not important here, it is important to note that care must be taken in the selection of appropriate quantum counters for lipofuscin work. Classically, RhodamineB has been used. However, its useful range extends only to 600 nm. This is very near the emission peak of many of the lipofuscin fluorophores, so a different quantum counter is required whose useful range extend beyond this point. Basic Blue 3, a laser dye, is useful as a quantum counter up to 700 nm (7, 28), and has been selected for spectral correction in our lipofuscin research. Thus appropriate instrumental calibration and correction procedures are exceedingly important in this work. Once properly calibrated, the measured spectra make perfect sense in terms of what one observes subjectively of the nature of the lipofuscin-related fluorescent emissions.

IMPLICATIONS FOR THE THEORY OF LIPOFUSCINOGENESIS

Based on results from spectrally corrected fluorescence emission analyses, it must be concluded that the fluorescent molecules accumulating in human RPE lipofuscin granules are not the blue emitting fluorophores described in the lipid autoxidation literature. Imine-conjugated Schiff base fluorescence cannot account for the RPE lipofuscin fluorescence.

Several questions arise from these conclusions. Do lysosomal enzymes modify the imine-conjugated Schiff base linkage through further polymerization or rearrangement? Is an altogether different chemistry involved? Others have invoked vitamin A metabolites, or low molecular weight proteinaceous fluorescence influenced by microenvironmental factors to explain the fluorescence of a theoretically related lipopigment, ceroid (35, 44).

CURRENT PROGRESS

It has become exceedingly important to identify the molecular structure of the fluorophores contained within the lipofuscin granules in order to reevaluate their source(s) of origin, mechanisms of accumulation, and possibilities for removal or manipulation. The first step in this process is the extraction, separation and purification. Previous attempts have failed to separate any of these fluorophores (19, 43). We have made progress in this regard and have made some significant observations in the process.

A thin-layer chromatographic system has been developed for the initial separation. At least ten fluorescent fractions are separable from RPE lipofuscin. Not only is this technique useful as a first step in purification, but it also has served as a useful method for generating a fluorophore fingerprint of chloroform-soluble fluorophores from different tissues and different species.

Using this technique, several points have become evident:

All of the fluorophores isolated from whole RPE extracts are present in extracts of isolated RPE lipofuscin granules.

Ten fluorescent fractions have now been separated from human RPE lipofuscin (11). They fall into four categories based upon their fluorescent excitation/emission characteristics: a pair of green-emitting fractions (330 nm Ex; 520 nm Em) possibly related to vitamin A; three yellow/green-emitting fractions (280, 330 nm Ex; 568 nm Em); one golden yellow-emitting fraction (280, 330 nm Ex;585 nm Em); and four orange/red emitting fractions (285, 335, 415 nm Ex; 605,633, 670 nm Ex). The existence of these well defined, chromatographically separable fluorophores also contradicts the notion that the lipofuscin granule contents are a hopelessly crosslinked, heterogeneous mixture of digestive byproducts.

None of the RPE lipofuscin fluorophores are similar to either the ICSB fluorophores or polymerized malonaldehyde (9), the latter of which has also been put forward as the material accumulating in lipofuscin (39). In fact, no blue-emitting fluorophores are present in human RPE lipofuscin granule extracts.

From a comparative stand point, human heart lipofuscin fluorophores which are extractable into chloroform differ in chromatographic mobility, but not in qualitative fluorescent emission properties from the fluorophores of human RPE (9). The human RPE fluorophores are essentially identical to aged rat RPE fluorophores (24). Only some of the human RPE fluorophores appear in aged cat RPE, and they differ from fluorophores isolable from the RPE of cats affected by Chediak-Higashi syndrome (5).

Work is now progressing toward the further purification, spectral analysis and molecular identity of the fluorescent molecules of human RPE lipofuscin (10, 12). When these compounds are identified, and their origin is understood, the theory of lipofuscinogenesis may need to be modified. It may be possible to design substances capable of manipulating in vivo RPE lipofuscin levels. The role, if any, in the etiology of AMD may then be elucidated. If light proves to be a factor in lipofuscin accumulation, and if lipofuscin accumulation, in turn, proves to be a factor in AMD , then action spectra can be determined and we can begin to redesign the lighting environment so as to reduce the prevalence of AMD and allow optimal vision in the elderly. Many basic questions need to be answered in the continuing research effort. In the near future, it is hoped that a first step can be made by identifying the compounds which accumulate within the lipofuscin granules of the RPE.

ACKNOWLEDGEMENTS

This research has been supported by grants from the National Institutes of Health (EY-06458), Research to Prevent Blindness, and the Children's Brain Diseases Foundation.

1. Anderson RE and Maude MB, 1970, Phospholipids of bovine rod outer segments. Biochemistry 9:3624-3628.

2. Besharse JC, 1982, The daily light-dark cycle and rhythmic metabolism in the photoreceptor-pigment epithelial complex. In: Progress in Retinal Research, Volume 1. Osborne NN and Chader GJ, editors, Pergamon Press, New York, pp. 81-124.

3. Burns RP and Feeney-Burns L, 1980, Clinico-morphologic correlations of drusen of Bruch's membrane. Tr Am Ophth Soc 78:206-225.

4. Chio KS and Tappel AL, 1969, Synthesis and characterization of the fluorescent products derived from malonaldehyde and amino acids. Biochemistry 8:2821-2827.

5. Collier LL, Eldred GE, King EJ, and Prieur DJ, 1986, Fluorophores of Chediak-Higashi feline retinal pigment epithelium. Invest Ophthalmol Vis Sci (Suppl) 27:101.

6. Delmelle M, 1977, Retinal damage by light: Possible implication of singlet oxygen. Biophys Struct Mechanism 3:195-198.

7. Demas JN, Pearson TDL, and Cetron EJ, 1985, Luminescence quantum counter comparator and calibration of new quantum counters for light intensity measurements. Anal Chem 57:51-55.

8. Desai ID and Tappel AL, 1963, Damage to proteins by peroxidized lipids. J Lipid Res 4:204-207.

9. Eldred GE, 1987, Questioning the nature of the fluorophores in age pigments. In: Advances in Age Pigments Research. Totaro EA, Glees P, and Pisanti FA, editors. Pergamon Press, Oxford, pp. 23-36.

10. Eldred GE, 1988, Vitamins A and E in RPE lipofuscin formation and implications for age-related macular degeneration. In: Inherited and Environmentally Induced Retinal Degenerations. Lavail MM, editor. Alan R. Liss, Inc., New York, (in press).

11. Eldred GE and Katz ML, 1988, Fluorophores of the human retinal pigment epithelium: Separation and corrected spectral characterization. Exp Eye Res 47:71-86.

12. Eldred GE and Katz ML, 1988, Possible mechanism for lipofuscinogenesis in the retinal pigment epithelium and other tissues. In: Lipofuscin---1987: State of the Art. Zs.-Nagy I, editor. Akademiai Kiado and Elsevier Science Publishers, Budapest and Amsterdam, pp. 185-211.

13. Eldred GE, Miller GV, Stark WS, and Feeney-Burns L, 1982, Lipofuscin: Resolution of discrepant fluorescence data. Science 216:757-759.

14. Feeney L, 1978, Lipofuscin and melanin of human retinal pigment epithelium. Invest Ophthalmol Visual Sci 17:583-600.

15. Feeney-Burns L, Berman ER and Rothman H, 1980, Lipofuscin of human retinal pigment epithelium. Am J Ophthalmol 90:783-791.

16. Feeney-Burns L and Ellersieck MR, 1985, Age-related changes in the ultrastructure of Bruch's membrane. Am. J. Ophthalmol 100:686-697.

17. Feeney-Burns L, Hildebrand ES, and Eldridge S, 1984, Aging human RPE: Morphometric analysis of macular, equatorial, and peripheral cells. Invest Ophthalmol Vis Sci 25:195-200.

18. Ganley JP and Roberts J, 1983, Eye conditions and related need for medical care among persons 1-74 years of age: United States, 1971-72. DHHS Publ No (PHS) 83-1678.

19. Goebel HH, Zeman W, Patel VK, Pullarkat RK, and Lenard HG, 1979, On the ultrastructural diversity and essence of residual bodies in neuronal ceroid-lipofuscinosis. Mech Ageing Devel 10:53-70.

20. Hack MH, 1981, Lipofuscin: Another view. In: Age Pigments.Sohal RS, ed. Elsevier/North-Holland Biomedical Press, New York, pp 383-388.

21. Hack MH and Helmy FM, 1983, The melanins and lipofuscin. Comp Biochem Physiol 76B:399-407.

22. Ham WT Jr, Mueller HA, Ruffolo JJ Jr, Millen JE, Cleary SF, Guerry RK, and Guerry D III, 1984, Basic mechanisms underlying the production of photochemical lesions in the mammalian retina. Curr Eye Res 3:165-174.

23. Handelman GJ and Dratz EA, 1986, The role of antioxidants in the retina and retinal pigment epithelium and the nature of prooxidant-induced damage. In: Advances in Free Radicals in Biology and Medicine, Volume 2. Pryor WA, editor. Pergamon Press, New York, pp. 1-89.

24. Katz ML, Drea CM, Eldred GE, Hess HH, and Robison WG Jr, 1986, Influence of early photoreceptor degeneration on lipofuscin in the retinal pigment eqithelium. Exp Eye Res 43:561-573.

25. Katz ML, Robison WG Jr, Herrmann RK, Groome AB, and Bieri JG, 1984, Lipofuscin accumulation resulting from senescence and vitamin E deficiency: Spectral properties and tissue distribution. Mech Age Dev 25:149-159.

26. Klenk E, 1972, The polyenoic fatty acid and the problem of essential fatty acids. In: Current Trends in the Biochemistry of Lipids. Ganguly J and Smellie RMS, editors. Academic Press, New York, pp. 41-48.

27. Kline D, Sekuler R, and Dismukes K, 1982, Social issues, human needs, and opportunities for research on the effects of age on vision: An overview. In: Aging and Human Visual Function. Sekuler R, Kline D, and Dismukes K, editiors. Alan R. Liss, Inc., New York, pp. 3-6.

28. Kopf U and Heinz J, 1984, 2,7-Bis(diethylamino) phenazoxonium chloride as a quantum counter for emission measurements between 240 and 700 nm. Anal Chem 56:1931-1935.

29. Leibowitz HM, Krueger DE, Maunder LR, et al., 1980, The Framingham Eye Study Monograph: An ophthalmological and epidemiological study of cataract, glaucoma, diabetic retinopathy, macular degeneration, and visual acuity in a general population of 2631 adults. Surv Ophthalmol 24 (Suppl):335-610.

30. L'Esperance FA Jr, 1983, Ophthalmic Lasers: Photocoagulation, Photoradiation, and Surgery, Second Ed. C.V. Mosby Co., St. Louis, pp 170-171.

31. Lillie RD and Fullmer HM, 1976, Histopathologic Technic and Practical Histochemistry, Fourth Edition. McGraw-Hill Book Co., New York, p. 12.

32. Malshet VG and Tappel AL, 1973, Fluorescent products of lipid peroxidation: I. Structural requirement for fluorescence in conjugated Schiff bases. Lipids 8:194-198.

33. Malshet VG and Tappel AL, 1974, Fluorescent products of lipid peroxidation: II. Methods for analysis and characterization. Lipids 9:328-332.

34. Marmor MF, 1982, Aging and the retina. In: Aging and Human Visual Function. Sekuler R, Kline D, and Dismukes K, editiors. Alan R. Liss, Inc., New York, pp. 59-78.

35. Palmer DN, Husbands DR, Winter PJ, Blunt JW, and Jolly RD, 1986, Ceroid lipofuscinosis in sheep. I. Bis(monoacylglycero) phosphate, dolichol, ubiquinone, phospholipids, fatty acids, and fluorescence in liver lipopigment lipids. J Biol Chem 261:1766-1772.

36. Sarks SH, 1976, Aging and degeneration in the macular region. A clinicopathologic study. Br J Ophthalmol 60:324-341.

37. Seliger HH, 1978, Excited states and absolute calibrations in bioluminescence. Meth Enzymol 57:560-600.

38. Shimasaki H, Privett OS, and Hara I, 1977, Studies of the fluorescent products of lipid oxidation in aqueous emulsion with glycine and on the surface of silica gel. J Am Oil Chemist's Soc 54:119-123.

39. Siakotos AN and Munkres KD, 1981, Purification and properties of age pigments. In: Age Pigments. Sohal RS, editor. Elsevier/North-Holland Biomedical Press, New York, pp. 181-202.

40. Silverstein RM, Bassler GC, and Morrill TC, 1974, Spectrometric Identification of Organic Compounds, Third Edition. John Wiley & Sons, New York, pp. 231-257.

41. Tappel AL, 1975, Lipid peroxidation and fluorescent molecular damage to membranes. In: Pathobiology of Cell Membranes, Volume 1. Trump BF and Arstila AV, editors. Academic Press, New York, pp. 145-173.

42. Tappel AL, 1980, Measurement of and protection from in vivo lipid peroxidation. In: Free Radicals in Biology, Volume IV. Pryor WA, editor. Academic Press, Inc., New York, pp. 2-48.

43. Taubold RD, Siakotos AN, and Perkins EG, 1975, Studies on chemical nature of lipofuscin (age pigments) isolated from normal human brain. Lipids, 10:383-390.

44. Wolfe LS, Ng Ying Kin NMK, Baker RR, Carpenter S, and Andermann F, 1977, Identification of retinoyl complexes as the auto fluorescent component of the neuronal storage material in Batten disease. Science 195:1360-1362.

45. Young RW, 1987, Pathophysiology of age-related macular degeneration. Surv Ophthalmol 31:291-306.

46. Young RW, 1988, Solar radiation and age-related macular degeneration. Surv Ophthalmol 32:252-269.

47. Young RW and Bok D, 1979, Metabolism of the retinal pigment epithelium. In: The Retinal Pigment Epithelium. Zinn KM and Marmor MF, editors. Harvard University Press, Cambridge, Massachusetts, pp. 103-123.

48. Zinn KM and Benjamin-Henkind JV, 1979, Anatomy of the human retinal pigment epithelium. In: The Retinal Pigment Epithelium. Zinn KM and Marmor MF, editors, Harvard University Press. Cambridge, Massachusetts, pp. 3-31.

UPDATE 1990

Since the writing of the original manuscript, significant progress has been made both in further demonstrating the differences between autofluorescent products of in vitro lipid oxidation and lipofuscin age pigment fluorophores, and toward the spectral characterization of the lipofuscin fluorophores of the human retinal pigment epithelium. The results of our work in this field have been published as follows:

Eldred, G.E., 1989, Vitamins A and E in RPE lipofuscin formation and implications for age-related macular degeneration. In: Inherited and Environmentally Induced Retinal Degenerations. LaVail MM, Anderson RE and Hollyfied JG (editors) Alan R. Liss, Inc., New York. ppl 113-129.

Eldred, G.E. and Katz, M.L., 1989, The autofluorescent products of lipid peroxidation may not be lipofuscin-like. Free Radical Biol. Med., 7:157-163.

Eldred, G.E. and Katz, M.L., 1991, The lipid peroxidation theory of lipofuscinogenesis cannot yet be confirmed. Free Radical Biol. Med., 10:(in press).

THE EFFECTS OF AGE ON PARALLEL PROCESSING IN THE HUMAN OPTIC NERVE

Alfredo A. Sadun, Michael Miao and Betty M. Johnson

USC School of Medicine, Dept. of Ophthalmology
Doheny Eye Foundation
1355 San Pablo St., Los Angeles, CA 90033

I. CLINICAL EVIDENCE OF PARALLEL PROCESSING

Parallel processing involves transmission for central processing of relatively independent packets of information along spatially separate pathways. As an analogy, one can consider a postman who carries separate packets of information in the form of different letters, each with its own address, each delivered separately. And yet, just as the mailman carries all his packets of letters in one pouch, the optic nerve must serve as a common conduit for all visual information transduced at the level of the eye and integrated in the brain.

There are many clinical examples that serve to illustrate that visual information can indeed be separated into distinct parameters which may be independent of each other. Diseases affecting one part of the eye can disproportionately affect one parameter of vision over another. For example, an injury to the optic nerve may have little impact on visual acuity, yet produce an afferent pupillary defect (50). Conversely, a small macular lesion will have an enormous impact on visual acuity and yet not produce any pupillary defect.

We investigated one parameter of vision that seemed to be markedly impaired by certain optic neuropathies: brightness-sense. A simple device employing pairs of cross polarizing filters, developed by Sadun and Lessell (38), enabled us to measure brightness-sense or, more specifically, the brightness sense disparity between a patient's two eyes (Fig. 1).

Subjects from three general populations were tested: twenty-five normal (control) subjects, thirty-four patients with ocular disorders other than those involving the optic nerve, and sixty-one patients with optic neuropathies. Patients with unilateral optic nerve lesions showed marked reductions in brightness-sense. Patients with retinal lesions, however, had only mild impairment of brightness-sense, and patients with unilateral cataracts or other media disorders showed no diminution of brightness-sense. It was apparent that patients with optic nerve diseases demonstrated a brightness-sense deficiency disproportionately more severe than the loss of visual acuity.

Knowing that different visual functions are differentially impaired by certain diseases raises an important question: are we

The Effects of Aging and Environment on Vision
Edited by D.A. Armstrong *et al.*, Plenum Press, New York, 1991

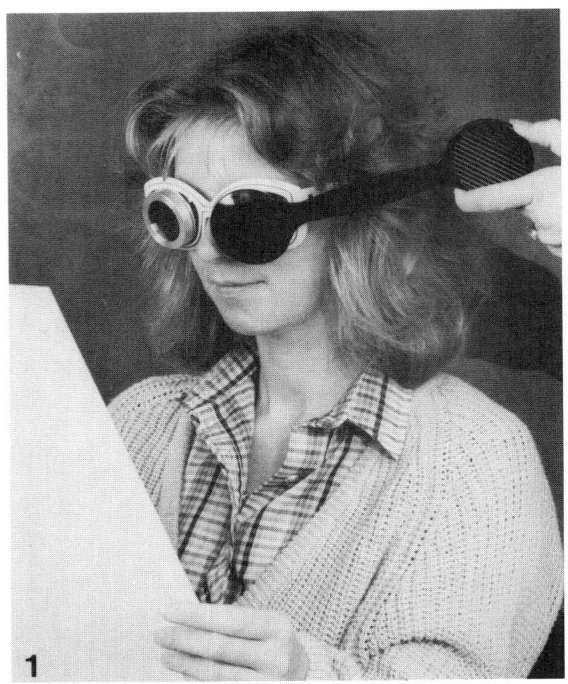

Figure 1. Subject wears device while regarding white paper under normal room illumination. Subject is asked to compare brightness of target as each eye is alternately occluded.

always testing the right visual functions? It would appear that, at least in certain circumstances, we are not. For example, Quigley et al. (34) performed post-mortem studies on optic nerves of patients who were suspected of having glaucoma. They showed that there could be loss of up to 50% of axons at the level of the optic nerve head without any clinical evidence of loss of visual function. In other words, patients with only half the normal complement of optic nerve fibers had normal visual acuity, normal color vision, and full visual fields. In measuring the possible effects of aging on vision, it will thus be necessary to be precise in defining separate visual parameters. It is likely that certain visual functions deteriorate more rapidly than do others in the process of aging. Identifying the most sensitive visual functions will help in developing the most sensitive clinical tests for measuring the potential impairments in vision seen with age (49).

II. SCIENTIFIC EVIDENCE FOR PARALLEL PROCESSING IN THE VISUAL SYSTEM

Animal studies have demonstrated that there exist physiologically distinct retinal ganglion cells. For example, retinal ganglion cells in cats have been shown to have X, Y, or W type responses (47); the bases for classifying these differing responses include the size and type of receptive field centers, linearity of the ON/OFF response, and axon conduction speed.

Similarly, monkey retinas have retinal ganglion cells that respond in X and Y-like manners (14, 36, 47). Human subjects have been studied by Enoch et al. (15), who demonstrated with quantitative layer-by-layer-perimetry that there are both sustained retinal responses and transient retinal responses. In addition, Weinstein (57) electrophysiologically recorded two types of retinal responses from isolated human retinas.

These data on retinal ganglion cell physiology correlate well with current anatomical data. Alpha, Beta, and Gamma retinal ganglion cells have been identified in the cat; each cell type has a unique cell body size, a distinct dendritic tree configuration, and a different axon size class (36, 47). It has been demonstrated in man that there may be classes of retinal ganglion cell axons distinguishable by size (39). Significantly, each axon size class seems to have a differential distribution to the several primary visual nuclei (39).

A third line of investigations lends support to the existence of parallel processing in the visual system. The common clinical teaching is that, in man, the retinal output is predominantly to the lateral geniculate nucleus and the remainder of the retinal projections is to the pretectum (51). However, a variety of other projections from the human retina have been demonstrated using the paraphenylene-diamine (PPD) staining method (Fig. 2). These retinofugal projections in man go to the lateral geniculate nucleus (LGN) (40, 42), pretectum (40, 42), superior colliculus (40, 42), pulvinar (40), and three hypothalamic nuclei: suprachiasmatic nucleus (40, 41), supraoptic nucleus (44), and paraventricular nucleus (40, 43). Thus, there appears to be in man several primary visual nuclei that separately process visual information in a parallel fashion. We have not yet delineated the precise visual functions subserved by each visual brain nucleus.

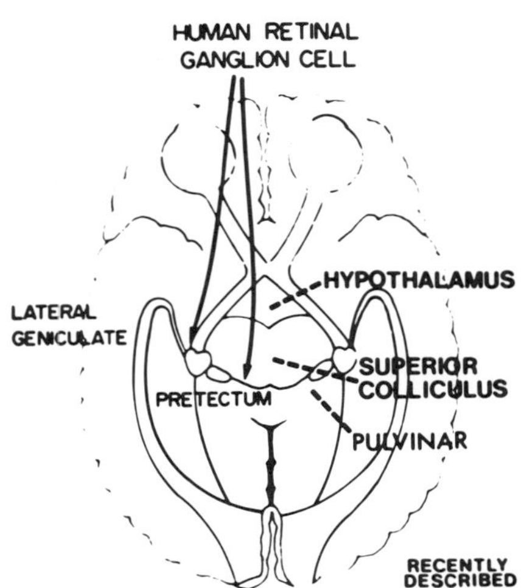

Figure 2. Schematic rendering of coronal section through the human brain, with selected retinofugal projections indicated. Recently described projections include the nuclei of the hypothalamus, the superior colliculus, and the pulvinar; the retinal projections to the lateral geniculate nucleus and the pretectum are well-known.

III. THE PPD TECHNIQUE AND VIDEO IMAGE MEASUREMENT SYSTEM

Until recently, very few neuroanatomical techniques have been successfully applied to study the human brain. However, a modification of an old myelin staining technique, using paraphenylene-diamine (PPD) (16, 45), has proven to work extremely well on human neural tissue (42). PPD chelates osmium, leaving a permanent, light opaque marker on cellular lipid elements. Therefore, the myelin sheaths of normal axons are intensely stained (Fig. 3a), as are the products of axonal degeneration (Fig. 3b) (37, 40, 42). The tissue is embedded in plastic, and semi-thin sections are stained with PPD. These thin sections permit high resolution light microscopy. By processing brain tissue as if for electron microscopy, subsequent ultrastructural confirmation of axonal degeneration is possible by examining the same or adjacent ultrathin sections (40).

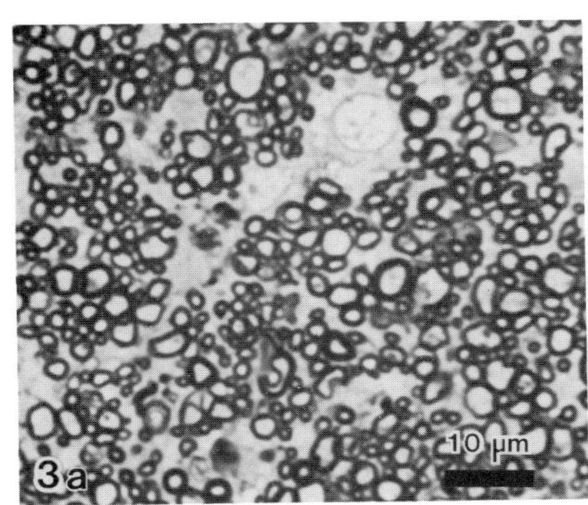

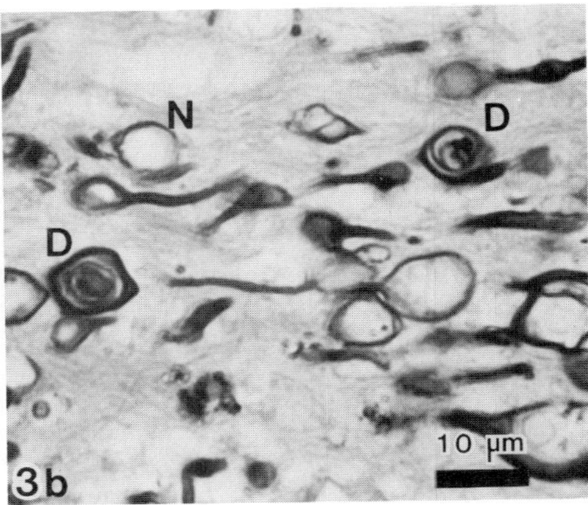

Figure 3a. Cross-section of normal human optic nerve shows tightly packed retinal cell fibers, which consist of dark myelin rings surrounding clear axoplasm (PPD, bar=10 um). **3b.** Cross-section of damaged human optic nerve (survival time=6 years). Both intact (N) and degenerated (D) fibers are seen (PPD, bar=10 um).

Fifteen optic nerves were obtained 5mm anterior to the optic chiasm from twelve brains at autopsy. Each optic nerve was sectioned, fixed in a glutaraldehyde/ paraformaldehyde buffered fixative, rinsed, osmicated, dehydrated by a series of ethanols, and embedded in plastic. Two micron sections were then cut on an ultramicrotome, placed on glass slides and stained in a one percent paraphenylene-diamine/methanol solution (13). The degree of staining intensity does not vary with the time of staining or rinsing, but rather with the thickness of the section.

Axon diameter measurements and axon population counts were obtained employing PPD-stained sections and a customized computer/microscope image enhancement and measurement system consisting of a video camera attached to a Zeiss standard-16 microscope, a high resolution monitor, and an enhanced Apple IIe computer (Fig. 4). Optic nerve cross-sections were viewed under the light microscope with a camera lucida attachment, divided into 20 sectors of approximately equal area, and assigned microscope stage coordinates (Fig. 5). The sectors so established were equivalent to those used by Quigley et al. (34). Four

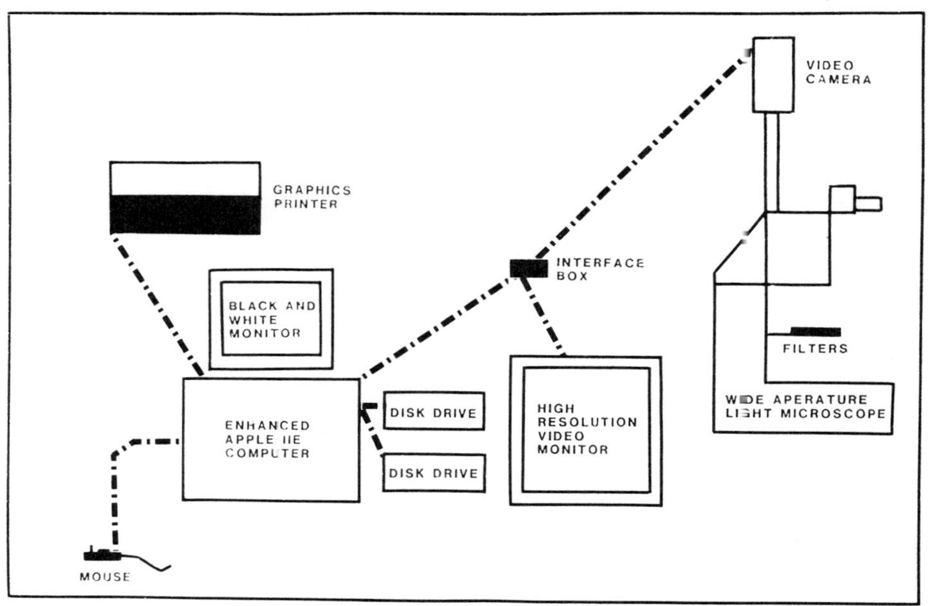

Figure 4. Schematic of video image measurement system (VIMS). VIMS consists of a video camera attached to a Zeiss light microscope, a high resolution black and white monitor, and an enhanced Apple IIe microcomputer with mouse.

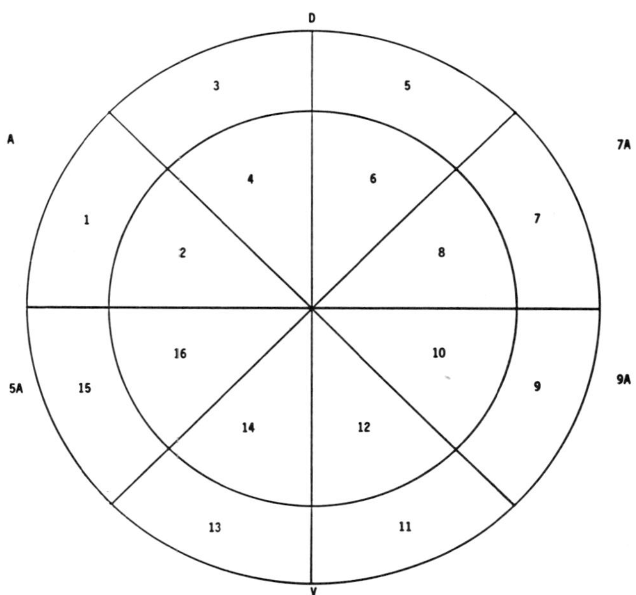

Figure 5. Circular diagram which is superimposed with a camera lucida over an image of an optic nerve, in order to assign sector coordinates. D=dorsal, V=ventral.

sectors were added laterally to account for the characteristic eliptical shape of the intracranial optic nerve (Fig. 6). Randomly selected fields were chosen by coordinates within each sector. From these fields, normal axons were hand-identified and measured (Fig. 7). The process was repeated until up to 1000

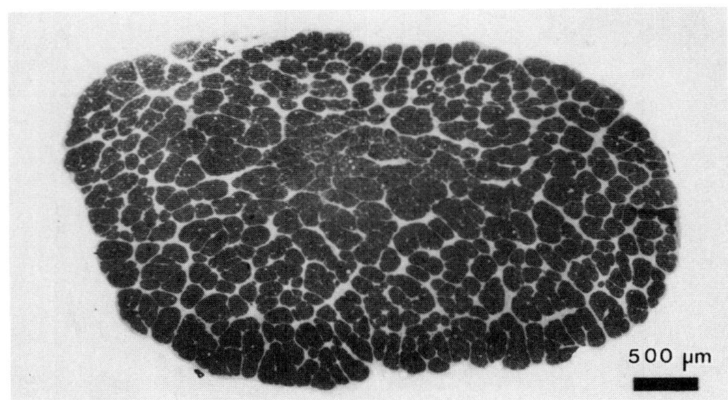

Figure 6. Low magnification micrograph of human intracranial optic nerve (7 mm anterior to chiasm); note characteristic eliptical shape (PPD, bar=500 um).

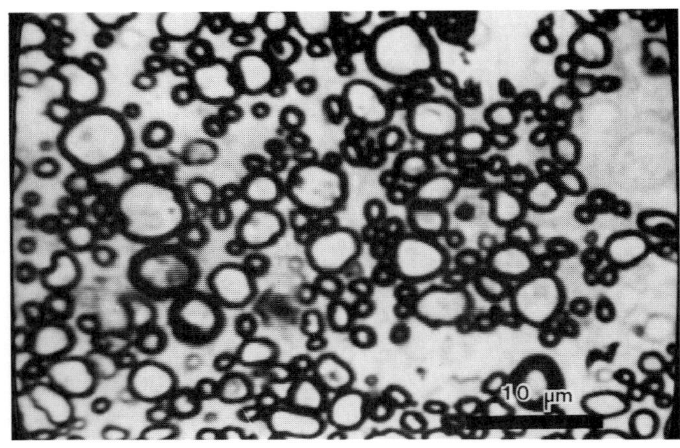

Figure 7. Normal human optic nerve as seen on monitor screen of video image measurement system. Normal axons are seen (PPD, bar=10 um).

axons were measured and counted from each of the 20 sectors. Employing a custom software package, the smallest internal diameter of each annulus was manually measured with a mouse device. The software automatically counts the axon and measures the diameter, area, and X versus Y location in the 1200 square micron measurement field. Degenerated axons are highly visible with this system (Fig. 8). A special wide aperature condensor arrangement, monochromatic short wavelength illumination, an electronic Fourier transform, and a high resolution monitor are important features of this system that permit good resolution even at a magnification of 5,600X.

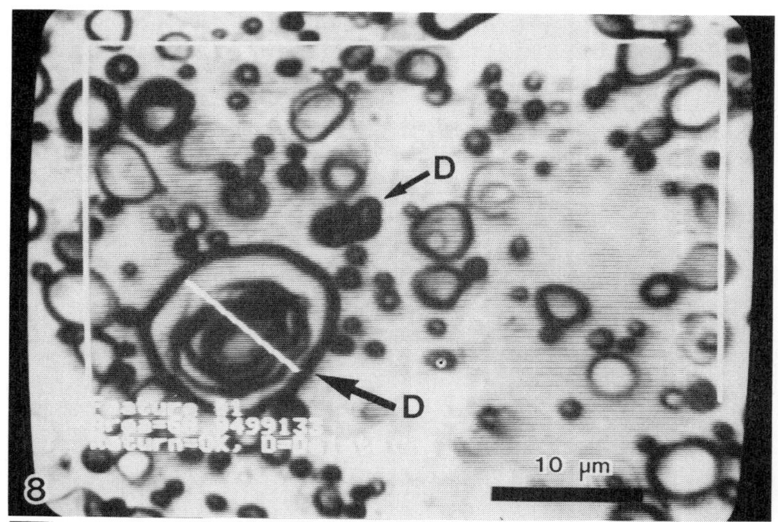

Figure 8. Damaged human optic nerve as seen on monitor screen of video image measurement system. A large degenerated axon has is visible, as are small degenerated axons (D) (PPD, bar=10 un).

IV. AGING CHANGES IN HUMAN VISUAL FUNCTION

A number of visual functions have been observed to deteriorate with age; the most thoroughly documented of these include changes in visual acuity (Snellen), color vision, depth perception, contrast sensitivity, and dark adaptation.

Visual acuities found in healthy young eyes tend to be considerably higher than those found in healthy older eyes (14, 18, 46). Collins and Britten (12) and others (9, 21, 23, 58) found the incidence of Snellen acuity of 20/20 or better in both eyes to increase from 6 to 20 years and to moderately decrease from 20 to 45, at which time the rate of decrease in acuity accelerates. Ferree et al. (17) showed that older observers required higher illuminances to achieve normal acuity than did younger observers.

Visual acuity losses occur in connection with a variety of age-related diseases and include an increased light scatter from cataract (14, 21, 25, 46, 53, 55), cellular loss in the retino-geniculo-striate pathway (28, 29, 54), senile macular degeneration, retinal pathology and open angle glaucoma (21, 23, 54).

Age-dependent declines in color vision after age 60 occur in red-green discrimination and for hues of blue and green (28). However, blue light is attenuated by the aging lens more than is light of longer wavelengths, resulting in a shift in the appearance of white toward yellow (11) and a relative darkening of blue-colored objects; there may also be slight reduction in the ability to discriminate between hues of closely related dominant wavelengths, particularly at low photopic light levels (11). Senile miosis and lenticular absorption and scatter can combine to elevate detection thresholds and to impair color recognition (11). Part of the decline in color vision may be associated with alterations of the dioptical mechanisms of the eye (28). Decreases in the discrimination of red-green and blue hues with age may be related to a loss of foveal cones, retinal bipolar and ganglion cells, and a loss of LGN neurons (28, 53).

Depth perception and stereopsis generally improve until middle-age. After 50, there is decline in stereopsis that is not fully attributable to losses in visual acuity. Cellular loss in the striate and extrastriate visual areas have been implicated in producing age-related impairments in stereopsis (28).

Contrast sensitivity is greatest between the ages of 20 and 30, and then decreases steadily to age 45 (6, 7, 29). More luminance is needed by the older person at all background levels (6, 29, 35, 59). Further increases in the requirement for great luminance disparity occurs at older ages (35).

Changes in contrast sensitivity that occur with age may be the result of glare from light scatter by the optical media of the eye (52). Some of the difference between contrast sensitivity in younger and older ages is due to senile miosis (52).

The Stiles-Crawford effect is reduced in elderly eyes, producing an increased sensitivity to stray light and a decreased sensitivity to direct light (16). Since the Stiles-Crawford effect predominantly involves the macular cones (28), this age-dependent change may affect ability to discern spatial details.

Dark adaptation significantly declines as age increases (56); this is most marked above the age of 60 (29). The target luminance must be doubled every 13 years to be above threshold in the dark-adapted eye (24). This increase in threshold is probably the result of pupillary (19) and neural changes (4, 5, 14, 19) as well as lenticular transmittance.

Thus with increasing age there is an increased incidence in ocular pathology, cataract formation and other media changes, senile miosis, changes in the Stiles-Crawford effect; retinal (especially macular) degenerative changes, and cortical neural processing deficits. These disorders produce impairments in visual acuity (Snellen), color vision, depth perception, contrast sensitivity, and dark adaptation. What morphological changes in the human optic nerve might underlie these impairments of vision?

V. OPTIC NERVE CHANGES WITH AGE

The human optic nerve, like other tissues in the body, is subject to changes with age. These changes, due to disease or degeneration, may negatively impact on vision. As the sole conduit of visual information from the eye to the brain, optic nerve functions embody all visual functions. There are many parameters of visual information that are mediated through the optic nerve before reaching various brain centers.

Attempts have been made to characterize the morphology of the optic nerve at different stages of the aging process and several characteristic changes have been observed: broadening of fibrous septae, production of corpora amylacea and lipofuscin, gliosis, and a diminution of axons (13).

In 1934, Arey & Schaible (2) studied fiber populations in two humans by means of silver and osmium stains. One individual's two optic nerves had counts of 761,000 and 697,000, respectively, whereas values of 1,287,000 and 1,318,000 were obtained from the other. No attempt was made to explain the different counts, to relate them to age (ages were not reported) or to assess variability in fiber populations.

Later work by Arey & Bickel (1) presented fiber population values from six additional human optic nerves. For the five individuals over 12 years, a mean of 1,304,306 fibers per optic nerve was calculated. The sixth individual was an 11-month-old infant with 1,096,500 optic nerve fibers; these counts were not significantly different from those reported by Arey and Schaible (2). Arey & Bickel also noted widely differing axon fiber sizes but did not attempt to quantitate axon populations by size.

Breusch & Arey (8) studied 12 human optic nerves under a magnification of 900X. They found a decrease in fiber population (1,200,000 to 871,000) corresponding to increases in age (47 to 71 years). They did not discuss the significance of the decreasing fiber counts with increasing age, perhaps because of the large range in individual values. Breusch and Arey calculated the mean fiber population to be 1,010,000.

Kupfer et al. (22) examined geographic differences in the optic nerve fiber densities and fiber populations in a 78-year-old individual. The authors reported using 1000X magnification and limited fiber diameter measurements to fibers 0.5 microns or greater. Fiber density values were not accompanied by an explanation of methodology, but our indirect conversion yields an estimated 100 to 300 fibers per 1000 square micron field. (For comparison purposes fiber densities will be presented in the standardized form of fibers per 1000 square micron field; it was sometimes necessary to compute and convert this from indirect information.) The center of the optic nerve exhibited a particularly high density of fibers (22). Two different counting techniques gave total axon fiber counts of 1,150,227 and 1,127,268, respectively. The authors did not identify the position along the optic nerve from which their tissue was obtained. However, inspection of their Figure 3 suggests that their specimen may have originated from the intracranial segment of the optic nerve.

Potts and colleagues (32) obtained <u>complete</u> counts of fiber populations in two young human optic nerves using an automated image analysis system; counts of 1,273,802 and 1,163,100 fibers/optic nerve were obtained on nerves from a 20- and 35-year-old, respectively. They also compared their optic nerve fiber counts to those of previous investigators: Polyak (30) and Oppel (27) gave fiber population values of slightly above one million axons per human optic nerve.

In a second article, Potts et al. (33) attempted to measure and categorize axons by size and to generate density maps of the optic nerve axons. Sections were obtained from tissue directly behind the globe. Some fibers were noted to have changed course, causing some axonal cross-sections to appear oblique instead of circular. These authors (33) stated that optic nerve axons do not normally have diameters larger than 2.5 microns, hypothesizing that 'artificially' large fibers may occur due to swelling resulting from trauma leading to the removal of the optic nerve (enucleation or death). The presence of larger fibers was not attributed to aging or disease processes. Fiber density maps showed 200 to 400 axons per 1000 square micron field. Potts et al. (33) combined all fibers of 1 micron diameter and smaller into one single group and then generated maps showing the various regions across the optic nerve that contained different percentages of small fibers. Since the magnification they described was very low, subdivisions within the range of 0.1 to 1.0 micron diameter were probably not possible. Nor is it likely that axons smaller than 0.5 microns in diameter were discernible. Since the great majority of normal axons are smaller than 1 micron in diameter (20, 39), Potts' fiber maps were not fully descriptive of the distribution of axon size classes. Their analysis of the fiber density maps and the small fiber maps led to the reiteration that optic nerves show tremendous individual variability (33).

Vrabec's (48) observations of the human optic nerve head also emphasized individual diversity in optic nerve appearance and axon number. Degenerative changes such as spherical swellings were seen most commonly at the periphery and were attributed to age.

Dolman et al. (13) examined several human optic nerves at different ages. Sections were obtained 1 to 2 mm behind the sclera. Poor myelin staining seen in the oldest optic nerves was attributed to loss of optic nerve fibers. Examination of optic nerves from individuals older than 60 years revealed diminished axon densities. Like Kupfer (22), Dolman et al. failed to provide a specific methodology for the axon density measurements. A rough conversion of their reported values gives 100 to 300 axons per 1000 square micron field. Increasing age and decreasing total optic nerve axon numbers were shown to be correlated, particularly for patients 60 years of age and older. They also noted fewer remaining retinal ganglion cells with age.

Quigley et al. (34) employed paraphenylene-diamine (PPD) to obtain counts of human optic nerve axon populations and to compare axon densities between regions. Their studies compared glaucomatous and normal human optic nerves. Optic nerve sections taken 2 to 3 mm behind the sclera were divided into 16 well-defined sectors of equal area and randomly sampled at 1000X magnification to count approximately 4% of the total axonal population. Measurements of the optic nerve area allowed for calculations of axon populations. The mean value for normal human optic nerves was 964,000 axons. Comparisons of different halves of the optic nerve showed that the upper, peripheral, and temporal halves had slightly more fibers than did their complementary sides. Their examination of glaucomatous human optic nerves showed marked axonal drop-out, often associated with minimal visual impairment. Although they did not address the issue of aging, Quigley et al. (34) developed an excellent methodology for optic nerve axon morphometry.

Employing the same protocol as Quigley et al. (34), Balazsi et al. (3) performed fiber counts on human optic nerves of different age groups. Tissue sections were taken 1 mm behind the globe and stained with PPD. They noted that the greater the time delay between patient death and optic nerve fixation the greater the amount of fiber loss and irregularity due to axonal swelling. Balazsi et al. attempted to correct for this fixation artifact through extrapolation, and concluded that the mean value for the 16 specimens (including several from older cases) should have been 1,200,000 axons/optic nerve. A multiple regression analysis was used to correlate age with axon population, resulting in an intercept value of 1,648,000 axons/optic nerve. They also calculated the rate of decay to be 5,637 axons/year, but the enormous scatter of their data points makes this value dubious. Balazsi et al. (3) suggested that axonal swelling and the failure to count smaller axons could have resulted in inaccuracies in previous axonal counts. The protocol employed by Balazsi et al. used 64 photomicrographs at 1715X magnification to sample 16 standardized sectors of approximately equal area. They sampled approximately 4 to 6% of the total neural area and counted about 10,000 axons. Balazsi et al. concluded that there was significant individual variation in fiber counts but projected that as many as 400,000 fibers may be lost to age in a 70 year life span. Their figures showing data on axon population as a function of age showed enormous scatter, reflecting either individual optic nerve variation or a high standard deviation due to methodology (3).

Our investigations have dealt with changes in four primary morphological features in the human intracranial optic nerve: total axonal population, axon density per sector, optic nerve area, and mean axon diameter. Using the previously described video image measurement system, these parameters were examined.

We saw a statistically significant decrease (t-test: $p<0.05$) in total axon population between a `young' age group (31-42 y.o.) and an `old' age group (60-86 y.o.). Axon populations ranged from 1,685,000 axons per nerve (42 years of age) to 759,000 axons per nerve (74 years of age). Mean axon diameters remained about 0.80 microns for all 20 sectors and in each age group. Mean axon density values ranged from 90 to 144 axons per 1000 square micron field; a qualitative, but not statistically significant, diminution was seen with increasing age.

Unlike Quigley et al. (34), we found no geographic differences in axon density when comparing dorsal, ventral, temporal, and nasal quadrants; nor were there significant or consistent differences between peripheral and central portions of the optic nerve. There were no statistical differences between the size class distributions of any sectors. The difference in retrobulbar versus intracranial optic nerve axon distributions is probably reflective of the repositioning of retinofugal fibers along 40 mm of optic nerve. Continuous reorganization of fiber packing probably occurs along the entire length of the optic nerve, optic chiasm, and the optic tracts (31). This constant reorganization of fiber order may represent persistent refinement of retinotopy. In addition, fibers may be repositioning and/or bifurcating prior to branching off into fascicles destined for various visual nuclei other than the lateral geniculate nucleus. Thus, comparison of fiber populations must be made with regard to position along the proximal/distal length of the optic nerve.

VI. DISCUSSION

Several previous studies of axon populations in the human optic nerve suffered from certain methodological deficiencies. Potts et al. (32) reported the magnification at which their measurements were made as 263X. This would make it impossible to accurately measure axons smaller than one micron in diameter, which would include most of the axons in the human optic nerve. Dolman et al. (13) did not define the criteria for describing subsets of axon populations in attempting to describe the heterogeneous organization of the optic nerve. Kupfer et al. (22) and Dolman et al. (13) did not define the size of the field upon which their fiber densities were based. The highest magnifications reported were those of Belazsi's, at 1715X (3). This also was suboptimal for accurate axonal counts and axonal measurements. 5600X magnification permitted measurement and counting of axons as small as 0.14 microns in diameter. Nonetheless we concur with Dolman et al. (13) that both total axon population and axon densities can decrease with aging without an apparent predilection for affecting specific axon size classes.

We found no differences in mean axon diameters with age. Therefore, axonal loss with aging probably affects axons of all sizes. Different axon size classes may mediate different visual functions. Ogden and Miller (26) and others have suggested that fiber diameter in the optic nerve is correlated to conduction velocity in processing various visual inputs. An impulse travels faster along a larger axon due to the cable properties of a regenerative depolarizing membrane. Thus with age, loss of axons from many classes may be reflected by impairment of many different visual functions. Testing the aging individual for changes in color vision, depth perception, contrast sensitivity, and dark adaptation may help clarify the many effects of aging on the human visual system.

What conclusions can we draw from these studies on aging changes in the human optic nerve? Scientifically, it is highly desirable to correlate decreased visual function with specific physiological changes in the visual system. It would be particularly advantageous to relate impairments of visual functions to loss of different retinal ganglion axon types within the optic nerve. From a clinical standpoint, it would be highly desirable to determine specific visual disabilities in the aged. Comprehensive visual testing may reveal these visual impairments.

Finally, it is important to consider the emotional duress imposed on a patient who has a visual impairment but is without a diagnosis. Establishing a clear organic basis for the patient's age-related visual dysfunction provides a basis for understanding and for offering a prognosis. There has existed an unfortunate tendency among some clinicians to dismiss complaints of visual impairment in the aged as diffuse and due to non-specific cerebral decompensations. We hope that some of the evidence described above will help focus attention on the fact that the aging process can bring about specific degenerative changes at the level of the optic nervehead, and that these changes may be the anatomical substrate upon which visual dysfunctions are based.

VII. SUMMARY

Parallel processing in the visual system has been described by clinical, physiological and anatomical studies in many species, including man.

Distinct retinal ganglion cell types have been demonstrated that respond differently depending upon several characteristics of the visual stimulus. These different retinal ganglion cells project their axons differentially to several primary visual nuclei. Clinical testing has shown disease-specific impairments of select visual functions.

Age correlated psychophysical changes have been described in the human visual system. Visual acuity, color vision, depth perception, contrast sensitivity, and dark adaptation have all been shown to decline in the elderly. Morphological changes in the human lens, vitreous, and retina have also been demonstrated. Age-related decreases in optic nerve fiber populations are found without selective loss of particular axon size classes, which suggests that aging broadly affects several retinal ganglion cell types and the several visual functions they subserve.

It is useful to define age-related changes in visual function as specific degenerative processes that have an anatomical basis. This is particularly important to the elderly, in whom visual deficiencies are often taken less seriously.

VII. REFERENCES

1. Arey, L.B., Bickel, W.H.,1935, The number of nerve fibers in the human optic nerve. Anat Rec 61 (suppl.):3.

2. Arey, L.B., Schaible, A.J., 1934, The nerve-fiber composition of the optic nerve. Anat Rec 58 (suppl.):3.

3. Balazsi, A.G., Rootman, J., Drance, S.M., Schulzer, M., Douglas, G.R. 1984, The effect of age on the nerve fiber population of the human optic nerve. Am J Ophthalmol 97:760-766.

4. Barlow, H.B., 1977, Retinal adaptation. Neurosci Res Prog Bull 15:394-416.

5. Barlow, H.B., Fitzhugh, R., Kuffler, S.W., 1957, Change of organization in the receptive fields of the cat's retina during dark adaptation. J Physiol 137:338-354.

6. Blackwell, O.M., Blackwell, H.R., 1971, Visual performance data for 156 normal observers of various ages. J Illum Eng Soc 1:3-13.

7. Bouma, P.J., 1936, The problem of glare in highway lighting. Phillips Tech Rev 1:225-229.

8. Bruesch, S.R., Arey, L.B., 1942, The number of myelinated and unmyelinated fibers in the optic nerve of vertebrates. J Comp Neurol 77:631-665.

9. Burg, A., 1966, Visual acuity as measured by dynamic and static tests. J Appl Psych 50:460-466.

10. Cane, V., Gregory, R.L., 1957, Noise and the visual threshold. Nature 180:1404-1405.

11. Carter, J.H., 1982, The effects of aging on selected visual functions: color vision, glare sensitivity, field of vision, and accommodation. Aging and Human Visual Function. Sekuler, R., Kline, D., Dismukes, K., eds.: 121-130. Alan R. Liss, Inc., New York.

12. Collins, S.D., Britten, R.H., 1924, Variation in eyesight at different ages, as determined by the Snellen test. Public Health Reports 39:3189-3194.

13. Dolman, C.L., McCormick, A.Q., Drance, S.M., 1980, Aging of the optic nerve. Arch Ophtalmol 98:2053-2058.

14. Ellenberger, C. Jr., 1984, Recent advances in the understanding of vision. Neuro-ophthalmol 4:185-206.

15. Enoch, J.M., 1978, Quantitative layer-by-layer perimetry. Proctor lecture. Invest Ophthalmol Vis Sci 17: 208-257.

16. Estable-Puig, J.F., Bauer, W.C. and Blumberg, J.M., 1965, Paraphenylenediamine staining of osmium-fixed, plastic-embedded tissue for light and phase microscopy. J Neuropathol Exp Neurol 24:531-535.

17. Feree, C.E., Rand, G., Lewis, E.F., 1934, The effect of increase of intensity of light on the visual acuity of presbyopic and non-presbyopic eyes. Trans III Eng Soc 29:296-313.

18. Frisen, L., Frisen, M., 1981, How good is normal visual acuity? A study of letter acuity thresholds as a function of age. Albrecht von Graefes Arch Klin Exp Ophthalmol 215:149-157.

19. Gunkel, R.D., Gouras, P., 1963, Changes in Scotopic visibility thresholds with age. Arch Ophthalmol 69:4-9.

20. Johnson, B.M., Sadun, A.A., 1985, Optic neuropathy: Correlation of visual field with patterns of anterograde degeneration. Invest Ophthalmol Vis Sci 26 (Suppl.):90.

21. Kornzweig, A.L., Feldstein, M., Schneider, J., 1957, The eye in old age. IV. Ocular survey of over one thousand aged persons with special reference to normal and distributed visual function. AM J Ophthalmol 44:29-37.

22. Kupfer, C., Chumbley, L., Downer, J.DeC, 1967, Quantitative histology of optic nerve, optic tract and lateral geniculate nucleus of man. J Anat 101:393-401.

23. Leibowitz, H.M., Krueger, D.E., Maunder, L.R., Milton, R.C., Kini, M.M., Kahn, H.A., Nickerson, R.J., Pool, J., Colton, T.L., Ganley, J.P., Loewenstein, J.I., Dawber, T.R., 1980, The Framingham Eye Study Monograph: An ophthalmological and epidemiological study of cataract, glaucoma, diabetic retinopathy, macular degeneration, and visual acuity in a general population of 2631 adults, 1973-75. Surv Ophthalmol 24:335-610.

24. McFarland, R.A., Domey, R.G., Warren, A.B., Ward, D.C., 1960, Dark adaptation as a function of age: I. A statistical analysis. J Gerontol 15:149-154.

25. Miline, J.S., Williamson, J., 1972, Visual acuity in older people. Gerontol Clin 14:249-256.

26. Ogden, T.E., Miller, R.F., 1966, Studies of the optic nerve of the rhesus monkey: nerve fiber spectrum and physiological properties. Vision res 6:485-506.

27. Oppel, O., 1963, Mikroskopische Untersuchungen uber die Anzahl und Kaliber der markhaltigen Nervenfasern im Fasciculus opticus des Menschen, Graefes arch. Opthalmol. 166:19.

28. Ordy, J.M., Brizzee, K.R., Johnson, H.A., 1982, Cellular alterations in visual pathways and the limbic system: Implications for vision and short-term memory. Aging and Human Visual Function. Sekuler, R., Kline, D., Dismukes, K., eds.:79-114. Alan R. Liss, Inc., New York.

29. Pitts, D.G., 1982, The effects of aging on selected visual functions: Dark adaptation, visual acuity, stereopsis, and brightness contrast. Aging and Human Visual Function. Sekuler, R., Kline, D., Dismukes, K., eds."131-159. Alan R. Liss, Inc., New York.

30. Polyak, S., 1941, Number of optic nerve fibers in man, in Polyak, S., ed.: The Retina. Chicago, 1941, University of Chicago Press. p. 447.

31. Polyak, S., 1957, The Vertebrate Visual System. Chicago: University of Chicago Press, pp. 301-309.

32. Potts, A.M., Hodges, D., Shelman, C.B., Fritz, K.J., Levy, N.S., Mangnall, Y., 1972, Morphology of the primate optic nerve. I. Method and total fiber count. Invest Ophthalmol 11:980-988.

33. Potts, A.M., Hodges, D., Shelman, C.B., Fritz, K.J., Levy, N.S., Mangnall, Y., 1972, Morphology of the primate optic nerve. II. Total fiber size distribution and fiber density distribution. Invest Ophthalmol 11:989-1003.

34. Quigley, H.A., Addicks, E.M., Green, W.R., 1982, Optic nerve damage in human glaucoma. III. Quantitative correlation of nerve fiber loss and visual field defect in glaucoma, ischemic neuropathy, papilledema, and toxic neuropathy. Arch Ophthalmol 100:135-146.

35. Richards, O.W., 1966, Vision at levels of night road illumination. XII. Changes of acuity and contrast sensitivity with age. Am J Optom Arch Am Acad Optom 43:313-319.

36. Rodieck, R.W., 1979, Visual pathways. Ann Rev Neurosci 2:193-225.

37. Sadun, A., 1975, Differential distribution of cortical terminations in the cat red nucleus. Brain Res 99:145-151.

38. Sadun, A.A., Lessell, S., 1985, Brightness-sense and optic nerve disease. Arch Ophthalmol 103:39-43.

39. Sadun, A.A., Schaechter, J.D., 1983, Retinal ganglion axon diameters and their differential distribution to the visual nuclei in man. Invest Ophthalmol Vis Sci 24 (Suppl):263.

40. Sadun, A.A., Schaechter, J.D., 1985, Tracin axons in the human brain: A method utilizing light and TEM techniques. JEMT 2:175-186.

41. Sadun, A.A., Schaechter, J.D., Smith, L.E.H., 1984, A retinohypothalamic pathway in man: Light mediation of circadian rhythms. Brain Res 302:371-377.

42. Sadun, A.A., Smith, L.E.H., Kenyon, K.R., 1983, Paraphenylenediamine: A new method for tracing human visual pathways. J Neuropathol Exp Neurol 42:200-206.

43. Schaechter, J.D., Sadun, A.A., 1985, A second hypothalamic nucleus receiving retinal input in man: The paraventricular nucleus. Brain Res 340:2430250.

44. Schaechter, J.D., Sadun, A.A., Johnson, B.M., 1985, A retinofugal pathway to the human supraoptic nucleus of the hypothalamus. Invest Opthalmol Vis Sci 26 (Suppl.):327.

45. Schultze, kW.H., 1917, Uber das paraphenylen-diamin in der histologischen Farbetechnik (katalytische Farbung) und uber eine neus Schnellfarbe methode der Nervenmarkscheinen am Gefrierschnitt. Sbl. allg. Pathol. Anat.: 28:257-260.

46. Sekuler, R., Kline, D., Dismukes, K., 1982, Introduction. Aging and Human Visual Function. Sekuler, R. Kline, D., Dismukes, K., eds.:117-120. Alan R. Liss, Inc., New York.

47. Stone, J., 1983, Parallel Processing in the Visual System: The Classification of Retinal Ganglion Cells and its Impact on the Neurobiology of Vision. Plenum Press, New York.

48. Vrabec, F., 1977, Age changes of the human optic nerve head: A neurophistologic study. Albrecht von Graefes Arch Klin Exp Opthalmol 202:231-236.

49. Wall, M., Sadun, A.A., 1986, Threshold amsler grid testing: Cross-polarizing lenses enhance yield. Arch Ophthalmol 104:520-523.

50. Walsh, F.B., Hoyt, W.F., 1969, Clinical Neuro-Ophthalmology. Vol. 1. 3rd Ed.. The Williams and Wilkins Company: Baltimore.

51. Walsh, F.B., Hoyt, W.F., 1969, Clinical Neuro-Ophthalmology. Vol. 3. 3rd Ed.. The Williams and Wilkins Company: Baltimore.

52. Weale, R.A., 1961, Retinal illumination and age. Trans Illum Eng Soc 26:95-100.

53. Weale, R.A., 1975, Senile changes in visual acuity. Trans Ophthalmol Soc UK 95:36-38.

55. Weale, R.A., 1978, The eye and aging. Gerontological Aspects of Eye Research. Hockwein, O., ed. Basel: Karger, 1-13.

56. Weale, R.A., 1982, Senile ocular changes, cell death, and vision. Aging and Human Visual Function. Sekuler, R., Kline, D., Dismukes, K., eds.:161-171. Alan R. Liss, Inc., New York.

57. Weinstein, G.W., Hobson, R.R., Baker, F.H., 1971, Extracellular recordings from human retinal ganglion cells. Science 171:1021-1022.

58. Weymouth, F.W., 1960, Effects of age on visual acuity. In Hirsch, M.J., Wick, R.E. (Eds.) "Vision of the Aging Patient." Philadelphia: Chilton Book Co., pp. 37-60.

59. Wolf, E., 1960, Glare and age. Arch Ophthalmol 64:502-514.

EVOKED POTENTIALS AND EEG SUGGEST CNS INHIBITORY DEFICITS IN AGING

Robert E. Dustman[1,2,3], Donald E. Shearer[1] and Rita Y. Emmerson[1,3]

[1] Neuropsychology Research, Veterans Administration Medical Center, Salt Lake City, UT
[2] Department of Neurology, School of Medicine, University of Utah Salt Lake City, UT
[3] Department of Psychology, University of Utah, Salt Lake City, UT

The central nervous system (CNS) comprises some fifty billion neurons. From the simplest reflex to the most complex activity, all behavior depends on the special ability of these neurons to communicate with the environment, with other neurons within the nervous system, and with muscles and glands. Communication between neurons is accomplished by electrochemical depolarization at nerve cell membranes that travels along cell processes to trigger the release of chemical neurotransmitters at synaptic clefts. These transmitters drift across the synaptic cleft and attach to receptor sites on the postsynaptic membrane effecting one of two possible changes: depolarization which biases membrane potentials of receiving neurons so that action potentials are more likely to occur (excitation), or hyperpolarization which decreases the likelihood of cell firing (inhibition). At any time, the probability of action potentials depends on the temporal and spatial summation of these excitatory and inhibitory effects. Thus, all behavior is controlled by excitatory and inhibitory neuronal activity and reflects an ongoing and dynamic shifting of the predominance of the two effects [55]. The importance of inhibition within the CNS is eloquently described by McGeer et al. [49, pg 133]. "We can think that inhibition is a sculpturing process. The inhibition, as it were, chisels away at the diffuse and rather amorphous mass of excitatory action and gives a more specific form to the neuronal performance at every stage of synaptic delay."

During aging the excitatory/inhibitory balance appears to be shifted in the direction of relatively less inhibition and consequently relatively more excitation [27, 31, 60]. An electrochemical bias towards greater excitation would be expected to affect behavior in subtle to more obvious ways depending upon the magnitude of inhibitory deficit. For example, reduced ability to inhibit irrelevant sensory and/or internal stimuli might result in impaired attention and concentration. Difficulty in suppressing or inhibiting an ongoing mental activity could contribute to increased behavioral rigidity or to diminished mental flexibility, an inability to quickly shift from one cognitive activity to another. Indeed, adult aging is associated with these kinds of cognitive impairments [7, 39, 58].

The Effects of Aging and Environment on Vision
Edited by D.A. Armstrong *et al.*, Plenum Press, New York, 1991

Enhanced neural excitation consequent to reduced inhibitory activity should also be reflected in the brain's more gross electrical manifestations, i.e., in evoked potentials (EPs) and EEG [27]. Thus, an inability of the older CNS to effectively suppress the effects of repetitious and relatively meaningless stimuli should result in enhances EP components; a gradual breakdown of inhibition could result in a greater homogeneity of electrical activity across the cortex. In the sections that follow, evoked potential and EEG findings which support the thesis of an excitation/inhibition imbalance in old age will be reviewed.

Underlying reasons for diminished inhibitory function in the elderly include cell loss in fairly specific cortical areas, reduced cerebral metabolic activity, and reduced function of neurotransmitters. Each has implications for neuronal inhibition. For example, relative to other sites, frontal association areas that have regulatory control over ascending excitatory pathways suffer excessive cell loss and experience reduced metabolic activity in old age [6,8,9,12,32,44], there is considerable thinning of horizontal dendrites that are presumed to facilitate inhibitory function [59], and activity of neurotransmitters believed to be involved in inhibition, e.g., the catecholamines, serotonin, and GABA, is reduced in old age [3,13,15,19,50,53].

Interestingly, adult aging appears to inversely parallel childhood development with respect to inhibition. As children mature they become less impulsive and are able to suppress or inhibit behaviors with less difficulty than at an earlier age [18,30,65,68]. The progressive improvement in inhibitory control during childhood undoubtedly reflects changes in neuronal and chemical systems that underlie excitatory-inhibitory relationships. Frontal association areas of children are not fully functional until adolescence as myelination of these areas is not complete until then [52,73], and there is some evidence from animal studies that the catecholamine and GABA systems continue to develop after birth [1,2]. Thus, children and the elderly appear to share CNS characteristics thought to be related to reduced central inhibition. As will be described later, the results from some electrophysiological measures that we interpret as reflecting inhibitory deficits are quite similar for children and the elderly.

MIDDLE-LATENCY VISUAL EVOKED POTENTIAL COMPONENTS

About two decades ago, Straumanis et al. [64] compared healthy young subjects, mean age of 24 years, with healthy oldsters whose mean age was 72 years on latency and amplitude of middle-latency (< 100 msec) visual evoked potential (VEP) components from occipital scalp. Middle-latency components are believed to reflect initial reception and processing of sensory information at visual cortex [26]. They found the middle-latency components not only occurred later in VEPs of the elderly but were of larger amplitude. The authors interpreted the larger amplitudes as indicating a reduction of central inhibition in old age [60,64]. Their findings were soon replicated in our laboratory [22,23]. VEPs were recorded from 215 normal subjects aged one month to 81 years. Recordings from occipital scalp, but not from more anterior areas, i.e., frontal and central scalp, demonstrated several obvious maturational changes (Figure 1). VEP components occurring after 100 msec were very large during childhood and rapidly diminished in size with development. In addition, we too observed that amplitude of VEP middle-latency components increased significantly during old age. As shown in Figures 1 and 2, the increase was visible as early as the fifth decade and became more accentuated with approaching senescence (the arrows in Figure 1 point toward a middle-latency component, P50). Age-related enhancement of middle-latency EP components has also been reported for auditory [71] and somatosensory stimulation (see below). The general consensus is that larger middle-latency EP components reflect weakened central inhibitory modulation of stimulus effects at sensory end organs and/or at sensory relay stations.

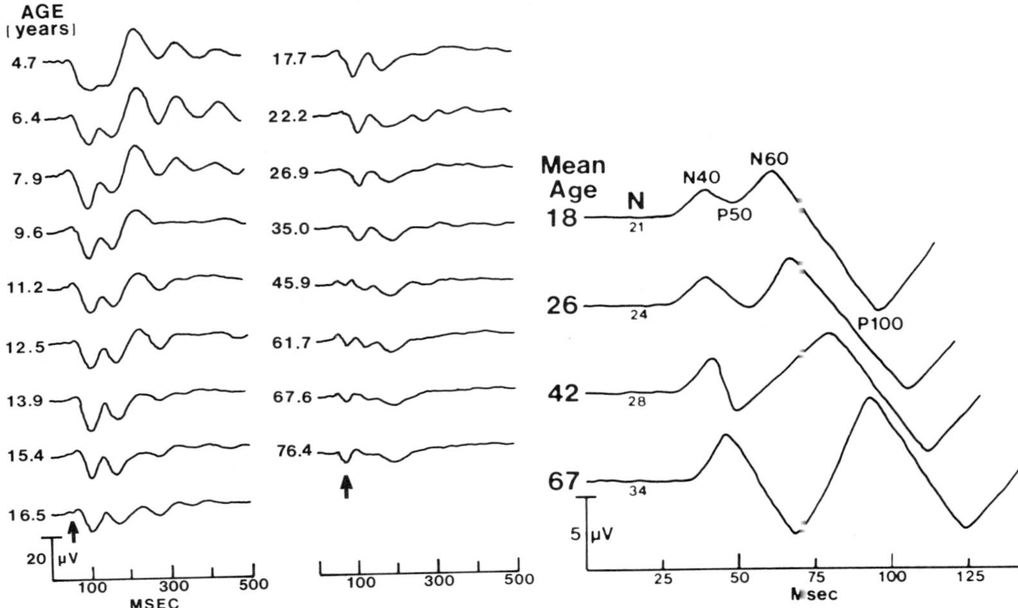

Figure 1. Age-related changes in VEPs. Shown are VEPs from 425 subjects aged 4-86 years. Each of the 17 age groups, designated by mean age at the left of the tracings, was composed of 25 individuals. The arrow beneath the last tracing in each column points to middle-latency components that become larger in occipital VEPs of older subjects.

Figure 2. VEPs from occipital scalp illustrating an age-related amplitude enhancement of middle-latency components. The composite VEPs were constructed from mean component latencies and amplitudes. The mean age and number of subjects (N) for each group are shown to the left of the tracings.

SOMATOSENSORY EVOKED POTENTIAL MIDDLE-LATENCY COMPONENTS

Somatosensory evoked potential (SEP) components occurring prior to 100 msec post-stimulus are also larger for old as compared to young adults. Shagass & Schwartz [61] stimulated the median nerve of 89 healthy subjects aged 15-80 years and observed that amplitudes of four components significantly increased with advancing age. Latencies of these components ranged from 20-70 msec.

Figure 3 illustrates SEPs from left central and left occipital scalp that were recently recorded in our laboratory from 60 right-handed healthy adults aged 21-79 years.

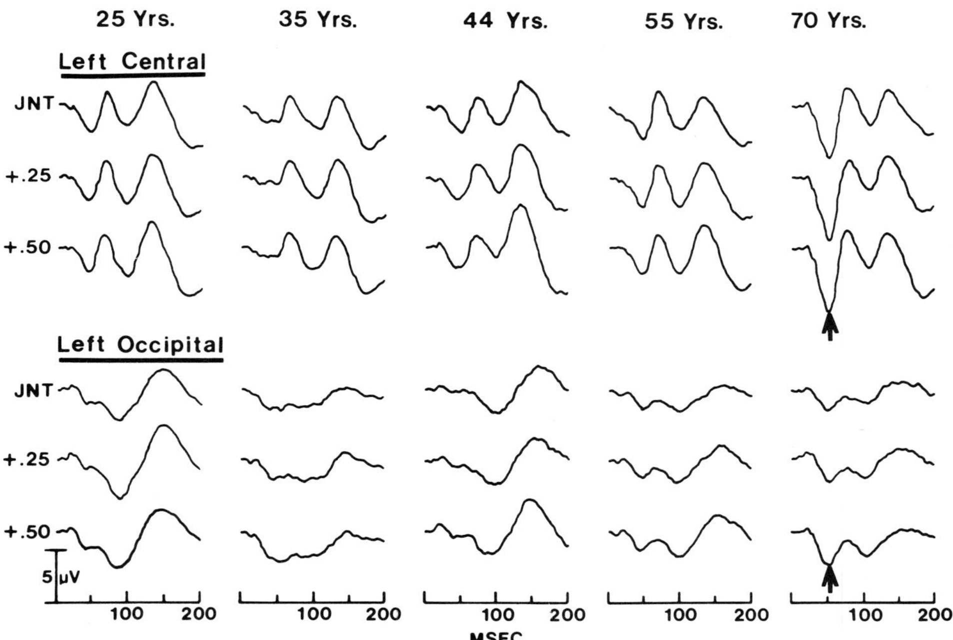

Figure 3. SEPs recorded from left central and left occipital scalp of 12 right-handed males in each of five age groups. Median nerve was stimulated with three intensities: that sufficient to elicit a just noticeable thumb twitch (JNT) and intensities 25% and 50% higher. Note that amplitude of the middle-latency wave complex, N30-P50-N70 (the arrows point to P50), is largest for the oldest subjects.

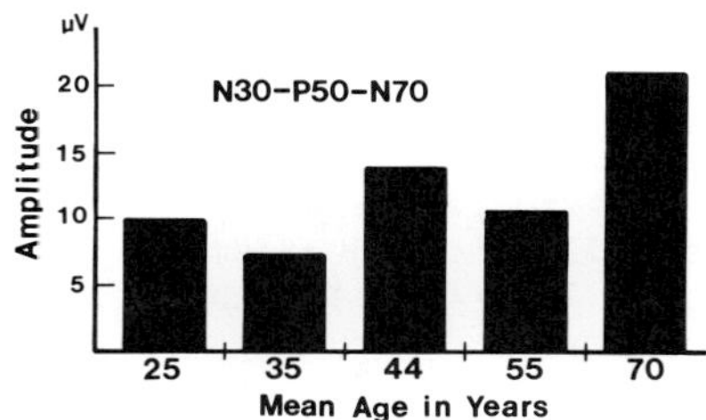

Figure 4. Mean amplitude of SEP middle-latency components N30-P50-N70 recorded from left central scalp (C3) of 12 right-handed males in each of five age groups. Note that the N30-P50-N70 complex was larger for the 70 year old subjects than for younger subjects. Amplitude for each subject was the sum of the amplitudes of the N30-P50 and P50-N70 components elicited by the strongest of three shock intensities, i.e., 50% above an intensity that produced a just noticeable thumb twitch.

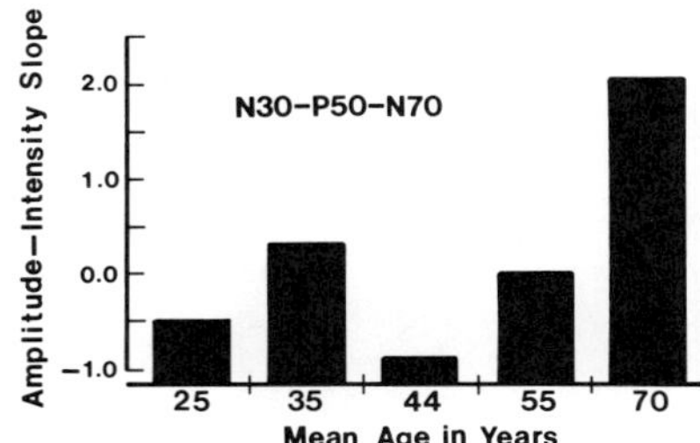

Figure 5. SEP amplitude-intensity (A/I) slope for 12 right-handed males in each of five age groups. A/I slope was computed on the summed amplitudes of the N30-P50- and P50-N70 components recorded from left central scalp (C3) and elicited by three intensities of shock. Note that the oldest subjects were the most sensitive to changes in shock intensity, i.e., A/I slope values were largest for these individuals.

Three intensities of median nerve stimulation were used: that necessary to produce a just noticeable thumb twitch (JNT) and intensities that were 25% and 50% higher. Although level of stimulation to produce a JNT was higher for older than for younger subjects, age differences for this measure did not reach significance. Age x Intensity ANOVAs revealed that amplitude of only one wave complex, N30-P50-N70, was differentially affected by age ($p< 0.02$). Mean amplitude for the oldest group of subjects was reliably larger than means for the younger groups (Figure 4).

A significant Age x Intensity interaction ($p < 0.02$) revealed that while shock intensity had no effect on N30-P50-N70 amplitude for the four younger groups, amplitude increased with higher shock values for the 70 year old subjects. This differential effect of age on SEP amplitude was evaluated by computing amplitude/intensity (A/I) slope values for each subject, i.e., the rate and direction of SEP amplitude change as a function of stimulus intensity [41,56]. While slope values for the four younger groups of subjects varied about zero (no effect of intensity on amplitude), a positive A/I slope was observed for the oldest subjects that was significantly greater than A/I slope for three of the remaining groups ($p < 0.02$) (Figure 5).

The lower half of Figure 3 illustrates SEPs recorded from left occipital scalp. Inspection of occipital SEPs reveals a definite N30-P50-N70 complex in SEPs of the oldest subjects (arrow points at P50) and, to a lesser degree, in SEPs of the next younger group (mean age 55). Similar to the findings for SEPs from central scalp. age had a significant effect on amplitude of these components ($p = 0.025$). N30-P50-N70 was larger for the 70 year old individuals than for the younger groups who did not differ with respect to N30-P50-N70 amplitude of this wave. These results parallel findings of Drechsler [20] and Liberson [47] who reported that while SEPs of young adults tended to be restricted to somatosensory recording sites, SEPs of old subjects were found to spread into other areas. Drechsler [20] attributed the wider distribution of SEPs to a reduction in central inhibitory function during old age. EEG alpha also shows an age related increase in cortical spread with the increase of spread accelerating during later years [33].

The above VEP and SEP findings show that old age is not only associated with "sensory receiving" evoked potential components that are larger than those for younger groups but also with increased homogeneity of brain electrical activity among recording sites. Both findings are compatible with an hypothesis of weakened inhibitory function in old age.

VEP COMPONENTS FROM FRONTAL AND CENTRAL SCALP

VEPs to three intensities of flash onset were recorded from 11 groups of 20 healthy males; mean ages of the groups ranged from 5 to 76 years [28]. Intensities were 1, 2, and 3 log units above a visual threshold that was determined for each subject [29]. Analysis of visual threshold data showed that subjects over 50 years of age required flash intensities significantly stronger than those used with younger subjects. Intensities for men 60 years and older were, on average, about 6.5 times as great as those for men aged 20-30 years [29]. This age differential in flash intensity agrees closely with values reported by Weale [67] for age changes in light transmission. Only about one-third as much white light reaches the retina of a 60 year old as reaches the retina of a young adult. The difference can be as large as a factor of 8-9 for a blue light (our stimulus flash is blue-white). Amplitudes of three VEP components, N90-P110, P110-N150, and N150-P200, recorded from left frontal (F3) and left and right central (C3, C4) scalp were measure (Figure 6). For each subject an overall amplitude measure was obtained by averaging across nine amplitudes; those of the three components in VEPs from each of the three scalp locations.

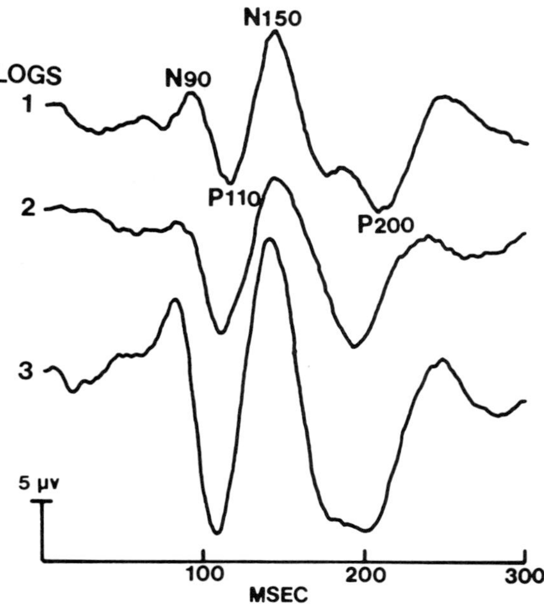

Figure 6. VEPs from central scalp elicited by three intensities of flash. Intensities were based on individually determined visual threshold and were 1, 2 and 3 $\log_{10}$ steps above threshold.

Figure 7 portrays age and intensity relationships for the overall amplitude measure. Two findings are of note. First, VEP amplitude for these frontal-central areas followed a U-shaped function with amplitude being larger for children and oldsters than for individuals of intermediate age. Second, in contrast to subjects in "prime of life" years, brighter flashes resulted in greater amplitude increments for the young and the old [28].

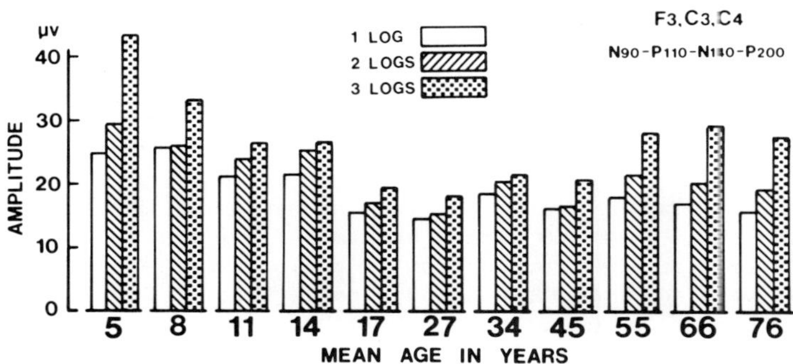

Figure 7. VEP amplitudes to three intensities of flash for 11 groups of 20 healthy males. Intensities were 1,2 and 3 $\log_{10}$ steps above individual visual thresholds. Amplitude was a composite measure of components N90-P110, P110-N140 and N140-P200 from VEPs recorded from left frontal (Fz) and left and right central(C3,C4)scalp. Note the U-shaped function across age and that the visual system of children and elderly was more responsive to bright flashes(3 logs)than for subjects of intermediate age.

For each subject, an A/I slope value was computed on the amplitudes elicited by the three intensities. A graph of the slope values across age revealed an impressive U-shaped curve similar to that reported above for overall amplitude. Slope values were largest for the young and the old (Figure 8) [28]. Inspection of Figure 8 reveals that significant age-related changes have occurred by the fifties. Others have also reported alterations in neurobiological function that have occurred by middle-age [69,70].

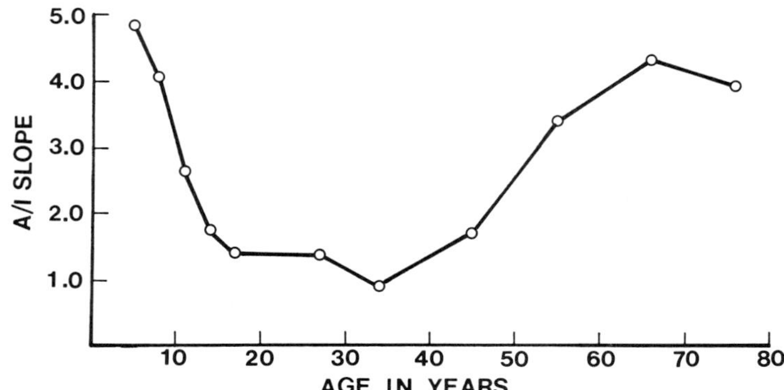

Figure 8. Mean amplitude-intensity (A/I) slope for 11 groups of 20 males. A/I slope followed a U-shaped function across age, values being larger for children and the elderly than for subjects between late adolescent to middle-age years.

We have since replicated the VEP intensity-amplitude findings associated with adult aging with a sample of five groups of 12 male subjects whose mean ages ranged from 25 to 70 years. Three flash intensities were again employed that were 1, 2 and 3 log steps above threshold values. In close agreement with our earlier study, mean threshold intensity for the 70 year old men was 6.2 times higher than the mean for the 25 year old group. An overall amplitude measure was obtained by averaging across nine amplitudes; those of N80-P105, P105-N140, and N140-P200 in VEPs from Fz (midline frontal), C3 and C4 recording sites. ANOVA computed on amplitudes elicited by the brightest flashes revealed a significant age effect ($F(4,55) = 3.68$, $p = 0.01$) with mean amplitudes increasing from the youngest to the oldest group. The slope of amplitude across flash intensity was computed for the combined amplitudes. Results showed a significant trend of larger A/I slope with increasing age ($F(4,55) = 2.56$, $p < 0.05$) (Figure 9). Our VEP findings, similar to those reported above for somatosensory stimulation, again suggest a relative inability of the old CNS to dampen cortical response to repetitive stimuli, particularly so for strong stimuli.

Validation for the thesis that damage to cortical association areas and deficits in neurotransmitter functioning may contribute to inhibitory loss and large amplitude VEPs is provided by studies of individuals with Down's syndrome (DS) and of alcoholics. Histologically, areas of the brain believed to exert inhibitory control over brain stem centers [59,63] are abnormally small in individuals with DS, showing generalized cellular agenesis and often both incomplete myelination ("dysmyelination") and early demyelination [4]. Abnormalities in monoamine neurotransmitter function of DS people have also been reported. Plasma levels of dopamine-B-hydroxylase, an enzyme which converts dopamine to norepinephrine are low [45], as are plasma levels of serotonin [14]. These transmitters are believed to play an inhibitory role in the nervous system by dampening reactivity to external and internal stimuli [19]. As shown in Figure 10, which portrays composite VEPs based on 66 subjects with DS and 66 age- and sex-matched normals, DS subjects have exceptionally large VEPs from frontal and central areas, but not from occipital sites [11,24]. Similar to normal elderly individuals, the visual system of people with DS responds to increased flash intensity with larger increments in VEP amplitude than occurs with normal controls [34].

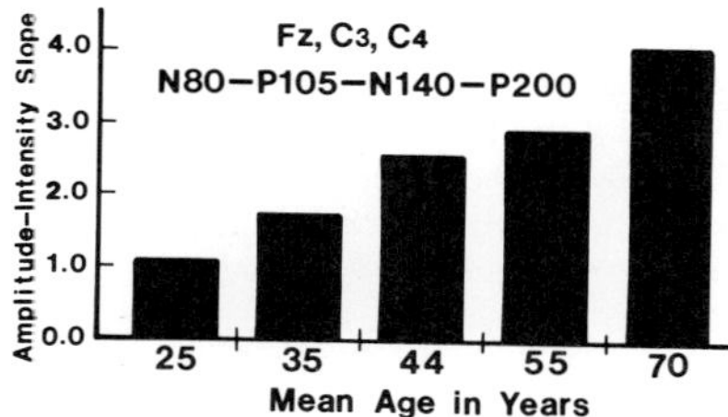

Figure 9. Mean amplitude-intensity (A/I) slope for five adult age groups. A/I slope increased in a fairly linear fashion across age. Intensities were 1. 2 and 3 $\log_{10}$ steps above individually determined visual thresholds. Amplitude was a composite measure that was the mean of VEP amplitudes N80-P105, P105-N140 and N140-P200 recorded from midline frontal (Fz) and left and right central (C3, C4) scalp.

Abnormally large VEPs and increased visual system sensitivity to bright flashes (enhanced A/I slope values) have also been noted for recently abstinent alcoholics; these were attributed to a decreased ability to inhibit neurophysiological activity [40,42]. Analyses of neurotransmitter enzymes and metabolites of withdrawing alcoholics have pointed to a reduced turnover of the monoamines, particularly of serotonin and dopamine [36,43]. In addition, alcoholism is associated with frontal cortical damage [17,46] which may contribute to an excitatory/inhibitory imbalance favoring excitation [see also, 21,51].

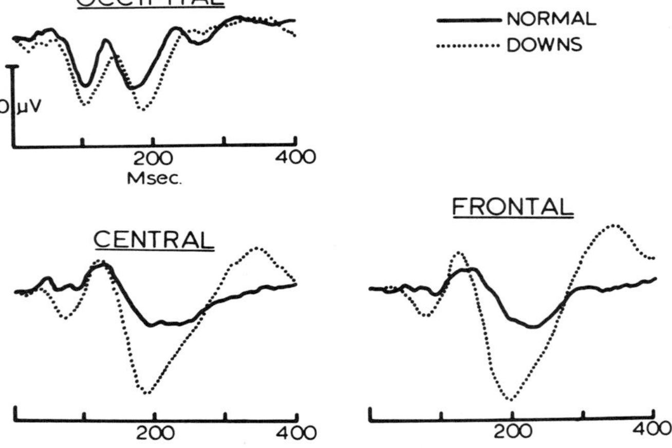

Figure 10. Composite VEPs recorded from 66 individuals with Down's syndrome (DS) and from 66 age-matched normal controls. Note that for DS subjects, VEPs from central and frontal scalp were much larger than VEPs from the controls; group differences for occipital VEPs were considerably smaller.

RELATIONSHIP OF AGE TO PATTERNED-UNPATTERNED VEP WAVEFORM SIMILARITY

Short duration flashes that project a checkerboard pattern evoke VEPs at occipital scalp that differ from VEPs elicited by unpatterned flashes [54]. Differences between "patterned" and "unpatterned" VEP waveforms are presumably related to inhibition as optimal detection of edges and contours is dependent upon an efficient inhibitory system [5,37,38,66]. Evidence that inhibitory processes are related to configuration of "patterned" VEPs is provided by stimulation of neurotransmitters believed to be inhibitory in function. Marcus [48] treated babies who had DS with 5-hydroxytryptophan (5-HTP) which is a precursor to serotonin (DS is associated with low levels of serotonin [14]). While their VEPs to patterned and unpatterned flashes were not different prior to 5-HTP, afterwards, the babies produced differentiated responses. In a second study, Schafer and McKean [57] stimulated the monoaminergic system of patients who had phenylketonuria, a disease associated with low catecholamine levels, by lowering phenylalanine level through dietary restriction or by the administration of indole and catecholamine precursors. After stimulation, patterned flash VEPs of the patients were clearly differentiated from VEPs to unpatterned flashes; such differentiation was not apparent before treatment.

We hypothesized that if central inhibition of healthy children and oldsters was impaired, the impairment should be reflected in patterned vs. unpatterned flash VEP comparisons, but only for VEPs from occipital scalp, i.e., areas overlying primary visual and visual association areas. Thus, waveforms of occipital VEPs to patterned and unpatterned flashes of the young and the old should demonstrate a higher degree of similarity than would be the case for adolescents and young adults whose visual systems would be expected to respond more efficiently to check contours. Patterned and unpatterned flash VEPs were recorded from 220 males aged 4-90 years [29]. For each subject digital values comprising the 0-300 msec segment of a patterned flash VEP were correlated with corresponding values for an unpatterned VEP. Correlations were computed for VEPs from occipital and central scalp areas. Figure 11 illustrates mean correlations for occipital VEPs. Coefficients of correlation followed a U-shaped curve across the life span, being significantly higher for young children and older adults than for subjects of intermediate ages. Note that the trend towards higher correlations in older age was noticeable by middle-age years. For recordings from central scalp there were no age-related variations in correlations of patterned vs. unpatterned flash VEP waveforms [29]. The greater similarity of occipital VEPs for the young and old subjects adds additional support to our thesis that inhibitory function is reduced during developmental and aging years.

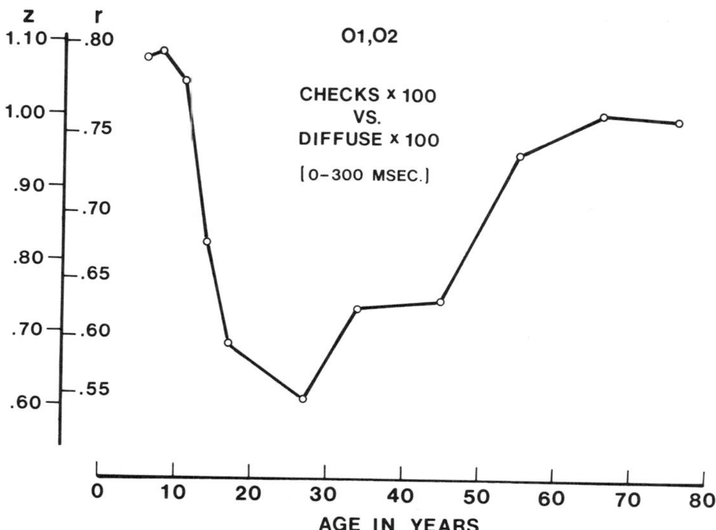

Figure 11. Life span changes in similarity of VEP wave forms elicited by patterned (checks) and unpatterned (diffuse) flashes. Each data point represents the mean correlation (r) and equivalent Fisher z-coefficient (z) obtained by correlating digital values comprising the 0-300 msec epoch of the two VEP waveforms. Flash intensity was 100 times (2 $\log_{10}$ steps above) visual threshold.

EEG: POWER SPECTRAL ANALYSIS AND CORTICAL COUPLING

We recorded three minutes of EEG from Fz, Cz, C3, Pz and Oz scalp areas of four groups of 20 healthy subjects while they sat awake and relaxed with eyes closed: males and females aged 25-35 years and males and females aged 55-70 years [25].

Power Spectral Analysis (PSA)

PSA was computed across a 3-15 Hz frequency band (.25 Hz resolution) on 40 artifact free four second segments of EEG using a fast Fourier approach [16,35]. A mean power loading for each 0.25 Hz frequency was obtained by averaging across 40 0.25 Hz values, i.e., those associated with the 40 four second-long EEG segments. Age and sex comparisons were made for four frequency bands: 5-7, 7-9, 9-11, and

11-13Hz. Power for each band was obtained by computing the mean of the eight values within the band. Analyses were done on $\log_{10}$ transforms of the band means.

Neither age nor sex had a significant effect on power loadings for any band or recording site. However, an Age x Area interaction was significant ($p< 0.001$) indicating differential power loadings for the young and old adults across recording areas. Power loadings were more variable across areas for the young than for the old (Figure 12). For all bands combined, the young subjects had larger power loadings at Cz and Pz but smaller loadings at Fz, C3 and Oz compared to the older people. Measures of power variability were obtained for each subject by computing the standard deviation of the loadings across the five electrode sites. This was done for each frequency band. Student t-test was employed to compare mean power variabilities (standard deviations) of the 40 young and 40 old subjects. Figure 12 shows the t-test results; for each band standard deviations were significantly smaller for the elderly ($p< 0.001$, $p< 0.001$, $p< 0.002$, and $p<0.002$ for the 5-7, 7-9, 9-11, and 11-13 Hz bands, respectively).

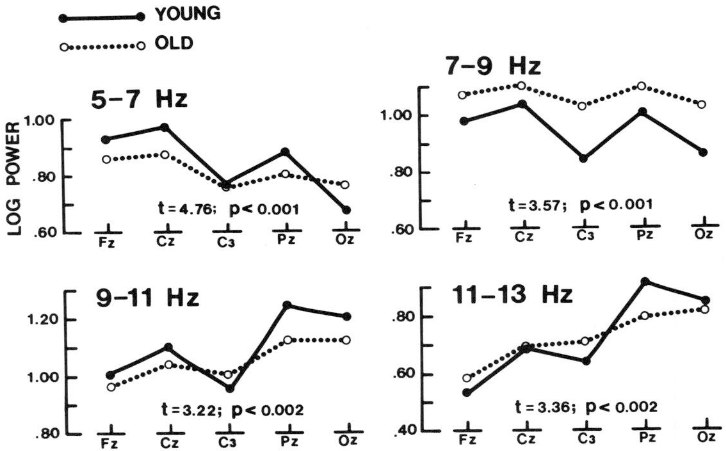

Figure 12. Mean $\log_{10}$ EEG power for forty subjects aged 25-35 years and forty aged 55-70 years illustrating reduced variability of power across recording sites for older subjects. Note that for each frequency band, interarea power variability was significantly less for the old subjects as shown by the t and p values. (Reproduced from Dustman, R.E., et al., ref. 25, with permission).

Cortical Coupling

Cortical coupling was computed to measure phase relationships between EEG patterns from pairs of electrodes [25]. This time-dependent measure is believed to reflect functional communication, i.e., information transmission among brain areas [10,72,74]. Our procedures, modeled after those of Yagi et al. [72], employed the Shannon-Weaver information transmission statistic Txy [62]. A 160 second epoch of EEG samples from each of two inputs (x and y) were classified as positive or negative and as larger or smaller than the previous sample. The x and y samples were each assigned to one of four categories: positive and rising, positive and falling, negative and rising, negative and falling. The correspondence between each pair of samples on these categories was entered as a tally in a 4 x 4 contingency table [25]. Txy was calculated from known probabilities associated with the table's cell, row and column totals. Cortical coupling (txy) values can range from 0.00 (no relationship between EEG patterns) to 2.00 (the EEG pattern recorded from one electrode site can be predicted from the pattern recorded from a second site). Cortical coupling values were computed for each possible pairing of the five electrodes (10 pairings).

An Age x Sex x Electrode pair ANOVA revealed highly significant effects on coupling values for Age (p< 0.001) and Electrode Pair (p< 0.001), but not for Sex (p> 0.20). Results of ANOVAs computed for individual electrode pairs are provided in Table 1. Note that coupling values for eight of the ten electrode pairs were significantly larger for the old than for the young subjects (p< 0.001).

Our findings of reduced variability of EEG power and higher cortical coupling values for the 55-70 year old individuals suggest there may be a breakdown of "functional autonomy" of areas within the older brain such that it response, at least electrically, in a more homogeneous or global manner. These results parallel results of other studies, i.e., an age-related increase in spread of SEPs from somatosensory areas [20,47] and an increased incidence of alpha some distance from occipital scalp [33].

Table 1

Means, Standard Deviations and ANOVA F and p Values for Cortical Coupling Data computed for 40 Young (25-35 Years) and 40 Older (55-70 Years) Subjects (ANOVA df= 1/76).

Electrode Pair	YOUNG ADULTS Mean	YOUNG ADULTS S.D.	OLDER ADULTS Mean	OLDER ADULTS S.D.	F	p
Fz-Cz	0.67	0.12	0.74	0.15	12.6	<0.001
Fz-C3	0.39	0.11	0.42	0.13	0.9	NS*
Fz-Pz	0.24	0.08	0.34	0.12	19.5	<0.001
Fz-Oz	0.09	0.04	0.16	0.08	26.3	<0.001
Cz-C3	0.54	0.13	0.58	0.15	1.7	NS
Cz-Pz	0.51	0.10	0.65	0.12	31.3	<0.001
Cz-Oz	0.16	0.06	0.28	0.10	41.9	<0.001
C3-Pz	0.40	0.09	0.52	0.11	29.3	<0.001
C3-Oz	0.15	0.05	0.29	0.12	46.5	<0.001
Pz-Oz	0.42	0.12	0.63	0.14	53.7	<0.001

*NS= Not Significant

ELECTROPHYSIOLOGICAL VS. NEUROPSYCHOLOGICAL MEASURES

We earlier noted that some behavioral changes that occur during adult aging may reflect a reduction in CNS inhibitory activity. If CNS inhibitory dysfunction underlies abnormalities in both the neuropsychological and electrophysiological expressions of human behavior, assessments of these two aspects of behavior should covary within individuals. We present results that suggest this may be true.

Data were obtained from the earlier described 60 male community volunteers aged 21-79 years. A measure of neuropsychological function was derived from the composite score from Shaie's test of Behavioral Rigidity [58], the Stroop Color Interference test [46], and the Trail Making Test [46]. These tests require ability to quickly shift mental set in order to score well and thus may involve inhibitory functioning. Three evoked potential measures that we and others believe to reflect inhibitory deficits provided an assessment of electrophysiological function: VEP A/I slope (frontal and central areas), SEP A/I slope, and amplitude of SEP component N30-P50. The three measures of neuropsychological and electrophysiological function were transformed to standard scores and means for the two sets of standard scores provided date points for each subject. The composite neuropsychological and electrophysiological measures were significantly correlated with each other ($r = 0.54$; $p < 0.001$) and with age (r's = 0.68 and 0.54, respectively; p's < 0.001). Even after removing variance accounted for by age, the electrophysiological and neurophysiological measures were significantly correlated ($r= 0.273$; $p < 0.036$). These results suggest that changes in the balance between CNS excitation and inhibition may underlie age-related changes in both electrophysiology and behavior.

In summary, results of the reviewed electrophysiological and behavioral studies are compatible with the hypothesis that central inhibitory strength weakens during adults aging. The relatively early age-related onset of change for some electrophysiological measures suggests that significant inhibitory losses have occurred by middle age.

ACKNOWLEDGEMENTS

The authors' research reported in this chapter was supported by the Veterans Administration

REFERENCES

1. Agrawal, H.C. and Himwich, W. A., 1970, Amino acids, proteins and monoamines of developing brain. In: Developmental Neurobiology, edited by W.A. Himwich, pp. 287-310. Thomas, Springfield.

2. Balzano, A.C. and Toffano, G., 1980, Ontogenetic development of GABA recognition sites in different brain areas. Pharmacol. Res. Commun., 12:495-500.

3. Beck, C.H.M., 1978, Functional implications of changes in the senescent brain: a review. Can. J. Neurol. Sci., 5:417-424.

4. Benda, C.E., 1969, Down's syndrome. Grune & Stratton, New York.

5. Bloom, F.E., Lazerson A., and Hofstadter, L., 1985, Brain, Mind, and Behavior. W.H. Freeman, New York.

6. Bondareff, W., 1977, The neural basis of aging. In: Handbook of the Psychology of Aging, edited by J.E. Birren and K.W. Schaie, pp. 157-176. Van Nostrand Reinhold, New York.

7. Botwinick J., 1973, Aging and Behavior. Springer, New York.

8. Brody, H., 1973, Aging of the vertebrate brain. In: Development and Aging in the Nervous System, edited by M. Rockstein, pp. 121-133. Academic Press, New York.

9. Bronson, G., 1965, The hierarchical organization of the central nervous system: implications for learning processes and critical periods in early development. Behav. Sci., 10:7-25.

10. Callaway, E. and Harris, P.R., 1974, Coupling between cortical potentials from different areas. Science, 183:873-875.

11. Callner, D.A., Dustman, R.E., Madsen, J.A., Schenkenberg, T. and Beck, E.C., 1978, Life span changes in the averaged evoked responses of Down's syndrome and nonretarded persons. Am. J. Ment. Def., 82:398-405.

12. Campbell, B.A., Lytle, L.D., and Fibiger, G.C., 1960, Ontogeny of adrenergic arousal and cholinergic inhibitory mechanisms in the rat. Science, 166:635-637.

13. Carlson, N.R., 1977, Physiology of Behavior. Allyn and Bacon, Boston.

14. Coleman, M. and Mahanand, D., 1973, Baseline serotonin levels in Down's syndrome patients. In: Serotonin in Down's syndrome, edited by M. Coleman, pp. 5-24. American Elsevier Publ. co., New York.

15. Concu, A., Carcassi, A.M., Piras, M.B., Blanco, S. and Argiolas, A., 1978, Evidence for a correlation between the latency of an early component of auditory evoked potentials and the brain levels of serotonin in albino rats. Experientia, 34:375-376.

16. Cooley, J.W. and Tukey, J.W., 1965, An algorithm for the machine computation of complex fourier series. Math Comput., 19:297-301.

17. Courville, C.B., 1955, Effects of Alcohol on the Nervous System of Man. San Lucas, Los Angeles.

18. Diamond, S., Blavin, R.S. and Diamond, F.R., 1963, Inhibition and Choice. Harper and Row, New York.

19. Douglas, W.W., 1978, Histamine and antihistamines: 5-hydroxytryptamine and antagonists. In: The Pharmacological Basis of Therapeutic, edited by L.S. Goodman and A. Gilman, pp. 590-629. Macmillan, New York.

20. Drechsler, F., 1978, Quantitative analysis of neurophysiological processes of the aging CNS. J. Neurol., 218:197-213.

21. Dustman, R.E., 1984, Alcoholism and aging: electrophysiological parallels. In: Alcoholism in the Elderly, edited by J.T. hartford and T. Samorajski, pp. 201-225. Raven, New York.

22. Dustman, R.E. and Beck, E.C., 1966, Visually evoked potentials: amplitude changes with age. Science, 151:1013-1015.

23. Dustman, R.E. and Beck, E.C., 1969, The effects of maturation and aging on the wave form of visually evoked potentials. Electroenceph. Clin. Neurophysiol, 26:2-11.

24. Dustman, R.E. and Callner, D.A., 1979, Cortical evoked responses and response decrement in nonretarded and Down's syndrome individuals. Am. J. Ment. Def., 83:391-397.

25. Dustman, R.E., LaMarche, J.A., Coh, N.B., Shearer, D.E. and Talone, J.M., 1985, Power spectral analysis and cortical coupling of EEG for young and old normal adults. Neurobiol. Aging, 6:193-198.

26. Dustman, R.E., Schenkenberg, T. and Beck, E.C., 1976, The development of the evoked response as a diagnostic and evaluative procedure. In: Developmental Psychophysiology of Mental Retardation, edited by R. Karrer, pp. 247-310. Thomas, Springfield.

27. Dustman, R.E. and Shearer, D.E., 1987, Electrophysiological evidence for central inhibitory deficits in old age. In: The London Symposia, (Electroenceph. Clin. Neurophysiol., Suppl. 39), edited by R.J. Ellingson, N.M.F. Murray and A.M. Halliday, pp. 408-412. Elsevier, Amsterdam.

28. Dustman, R.E. and Snyder, E.W., 1981, Life-span changes in visually evoked potentials at central scalp. Neurobiol. Aging, 2:303-308.

29. Dustman, R.E., Snyder, E.W. and Schlehuber, C.J., 1981, Life-span alterations in visually evoked potentials and inhibitory function. Neurobiol. Aging, 2:187-192.

30. Fishbein, H.D., 1976, Evolution, Development and Children's Learning. Goodyear, Pacific Palisades, CA.

31. Frolkis, V.V. and Bezrukov, V.V., 1979, Aging of the Central Nervous System. Karger, Basel.

32. Fuster, J.M., 1980, The Prefrontal Cortex. Raven, New York.

33. Gaches, J., 1960, Etude statistique sur les traces "alpha largement developpe" en fonction de l'age. La Presse Medicale, 68:1620-1622.

34. Gliddon, J.B., Busk, J. and Galbraith, G.C., 1975, Visual evoked responses as a function of light intensity in Down's syndrome and nonretarded subjects. Psychophysiology, 12:416-422.

35. Gold, G. and Rader, C.M., 1969, Digital Processing of Signals. McGraw-Hill, New York.

36. Gottfries, C.G., Knorring, L. von and Perris, C., 1976, Neurophysiological measures related to levels of 5-hydroxyindoleacetic acid, homovanillic acid and tryptophan in cerebrospinal fluid of psychiatric patients. Neuropsychobiology, 2:1-8.

37. Harter, M.R., 1971, Visually evoked cortical responses to the on- and off-set of patterned light in humans. Vision Res, 11:685-695.

38. Harter, M.R. and White, C.T., 1968, Effects of contour sharpness and check-size on visually evoked cortical potentials. Vision Res., 701-711.

39. Hoyer, W.J. and Plude, D.J., 1980, Attentional and perceptual processes in the study of cognitive aging. In: Aging in the 1980s, edited by L.W. Poon, pp. 227-238. American Psychological Association, Washington, D.C.

40. Knooring, L. von, 1976, Visual averaged evoked responses in patients suffering from alcoholism. Neuropsychobiology, 2:233-238.

41. Knorring, L. von, 1978, Visual averaged evoked responses in patients with bipolar affective disorders. Neuropsychobiology, 4:314-320.

42. Knorring, L. von and Oreland, L., 1978, Visual averaged evoked responses and platelet monoamine oxidase activity as an aid to identify a risk group for alcoholic abuse. A preliminary study. Prog. Neuro-Psychopharmac., 2:385-392.

43. Knorring, L. von and Perris, C., 1981, Biochemistry of the augmenting-reducing response in visual evoked potentials. Neuropsychobiology, 7:1-8.

44. Kuhl, D.E., Metter, E.J., Riege, W.H. and Hawkins, R.A., 1984, The effect of normal aging on patterns of local cerebral glucose utilization. Ann. Neurol., 15 (suppl.):S133-S137.

45. Lake, C.R., Ziegler, M.G., Coleman, M. and Kopin, I.J., 1979, Evaluation of the sympathetic nervous system in trisomy-21 (Down's syndrome). J. Psychiatr. Res., 15:1-6.

46. Lezak, M.D., 1983, Neuropsychological Assessment, 2nd edition. Oxford University Press, New York.

47. Liberson, W.T., 1976, Scalp distribution of somato-sensory evoked potentials and aging. Electromyogr. Clin. Neurophysiol., 16:221-224.

48. Marcus, M.M., 1970, The evoked cortical response: A technique for assessing development. Calif. Mental Health Res. Dig., 8:59-72.

49. McGeer, P.L., Eccles, J.C. and McGeer, E.G., 1978, Molecular Neurobiology of the Mannalian Brain. Plenum, New York.

50. McGeer, P.L. and McGeer, E.G., 1980, Chemistry of mood and emotion. Annu. Rev. Psychol., 273-307.

51. Porjesz, B. and Begleiter, H., 1981, Human evoked brain potentials and alcohol. Alcohol: Clin. Exper. Res., 5:304-317.

52. Rabinowicz, T., 1979, The differentiate maturation of the human cerebral cortex. In: Human Growth, vol. 3, edited by F. Falkner and J.M. Tanner, pp. 97-123. Plenum Press, New York.

53. Reader, T.A., 1978, The effects of dopamine, noradrenaline and serotonin in the visual cortex of the cat. Experientia, 34:1586-1588.

54. Regan, D., 1972, Evoked Potentials in Psychology, Sensory Physiology and Clinical Medicine. Chapman and Hall, London.

55. Roberts, E., 1972, Coordination between excitation and inhibition: development of the GABA system. In: Sleep and the Maturing Nervous System, edited by C.D. Clemente, D.P. Purpura and F.E. Mayer, pp. 79-98. Academic Press, New York.

56. Saaty, T.L., 1959, Mathematical Methods of Operations Research. McGraw-Hill, New York.

57. Schafer, E.W.P. and C.M. McKean, 1975, Evidence that monoamines influence human evoked potentials. Brain Res., 99:49-58.

58. Schaie, K.W., 1975, Test of Behavioral Rigidity (manual). Consulting Psychologists Press, Palo Alto, CA.

59. Scheibel, M.E. and Scheibel, A.B., 1975, Structural changes in the aging brain. In: Aging, vol. 1, edited by H. Brody, D. Harman and J.M. Ordy, pp. 11-37. Raven Press, New York.

60. Shagass, C., 1972, Evoked Brain Potentials in Psychiatry. Plenum Press, New York.

61. Shagass, C. and Schwartz, M., 1965, Age, personality and somatosensory evoked responses, Science. 148:1359-1361.

62. Shannon, C.E. and Weaver, W., 1949, The Mathematical Theory of Communication. University of Illinois Press, Urbana, IL.

63. Skinner, J.E. and Yingling, C.D., 1977, Central gating mechanisms that regulate event-related potentials and behavior. A neural model for attention. In: Progress in Clinical Neurophysiology, edited by J.E. Desmedt, pp. 28-68. Karger, Basel.

64. Straumanis, J.J., Shagass, C. and Schwartz, M., 1965, Visually evoked cerebral response changes associated with chronic brain syndromes and aging. J. Gerontol, 20:498-506.

65. Strommen, E.A., 1973, Verbal self-regulation in a children's game: impulsive errors on "Simon says". Child Develop., 44:849-853.

66. Uttley, A.M., 1979, Information Transmission in the Nervous System. Academic Press, London.

67. Weale, R.A., 1965, On the eye. In: Behavior, Aging and the Nervous System, edited by A.T. Welford and J.E. Birren, pp. 307-339. C.C. Thomas, Springfield, IL.

68. White, S.H., 1965, Evidence for a hierarchical arrangement of learning processes. In: Advances in Child Development and Behavior, edited by L.P. Lipsitt and C.C. Spiker, pp. 187-220. Academic Press, New York.

69. WoodruffpPak, D.S., Lavond, D.G., Logan, C.G. and Thompson, R.F., 1987, Classical conditioning in 3-, 30-, and 45-month old rabbits: behavioral learning and hippocampal unit activity. Neurobiol. Aging, 8:101-108.

70. Woodruff-Pak, D.S. and Thompson, R.F., 1988, Classical conditioning of the eyeblink response in the delay paradigm in adults aged 18-83 years. 3:219-229.

71. Woods, D.L. and Clayworth, C.C., 1986, Age-related changes in human middle latency auditory evoked potentials. Electroenceph. Clin. Neurophysiol., 65:297-303.

72. Yagi, A., Bali, L. and Callaway, E., 1976, Optimum parameters for the measurement of cortical coupling. Physiol. Psychol., 4:33-38.

73. Yakovlev, P.V. and Lecours, A.R., 1967, The myelogenetic cycles of regional maturation of the brain. In: Regional Development of the Brain in Early Life, edited by A. Minkowski, pp. 3-70. Davis, Philadelphia, PA.

74. Yingling, C.D., 1977, Lateralization of cortical coupling during complex verbal and spatial behaviors. In: Language and Hemispheric Specialization in Man: Cerebral ERPs, edited by J.E. Desmedt, pp. 151-160. Karger, Basel.

UPDATE 1990

We recently reported that CNS functioning may be enhanced by frequent participation in physical activity that contributes to cardiovascular health (1,2).

Our study sample comprised 30 men aged 21-31 years and 30 aged 50-62 years. All were in good health. Fifteen subjects in each age group had excellent-superior fitness levels; fitness levels of the others were poor-fair. Fitness level was documented by oxygen utilization (VO2max) during a maximal exercise test on a motorized treadmill. Measures of EEG, event-related potentials (ERPs), visual sensitivity and cognition were used to assess CNS functioning.

Expected age decrements were observed for most measures. Compared to the young men, performance of the older men was poorer on tests of visual sensitivity and cognition. ERP component latencies indicated that signal processing was slower for the older group. Measures of EEG waveform similarity across recording sites (cortical coupling) and of amplitude-intensity (A/I) slope of visually evoked potentials (VEPs), suggested that central inhibition was weaker for older than young men. (See our chapter for a description of these measures).

Significant fitness effects were also seen. Men who exercised frequently scored better on tests of visual sensitivity and cognition than those who had been sedentary. Fitness level was positively related to speed of signal processing as reflected by VEP latencies and to inhibitory function (VEP A/I slope and cortical coupling). Results for two measures, latency of the P300 component and cortical coupling, suggested that a lifestyle of physical activity can slow the rate of CNS decline. Older high fit men did not differ from the younger men on these measures while older sedentary men demonstrated age-related deficits.

We speculated the performance superiority of the physically active men was, at least in part, the result of more oxygen being available for cerebral metabolism.

References

1. Dustman R.E., Emmerson R.Y., Ruhling, R.O., Shearer, D.E., Steinhaus, L.A., Johnson, S.C., and Bonekat, H.W., 1990, Age and fitness effects on EEG, ERPs, visual sensitivity, and cognition. Neurobiol. Aging 11:193-200.

2. Emmerson, R.Y., Dustman, R.E., and Shearer, D.E., 1989, P3 latency and symbol digit performance correlations in aging. Exp. Aging Res., 15:151-159.

MACULAR DEGENERATION AND THE EFFECTS OF LIGHT

Michael F. Marmor

Division of Ophthalmology
Stanford University Medical Center
Stanford, California 94305

INTRODUCTION

Scholars have argued for years whether aging is inevitable. Do our tissues wear out by some biological timetable or is aging conditioned by environmental influences and chemical reactions that might be interrupted if we only knew how?

Is there really any distinction between "normal" aging that we all must anticipate and suffer, and pathologic aging that interferes with normal activity and that we should try to prevent? Although these distinctions are often blurred, they have practical significance in terms of patient care: The human lens yellows and hardens with age, but only if it loses enough clarity to be called a cataract will surgery be considered Age causes changes in the physical appearance of the retina, and in subtle aspects of visual perception, but it is a judgement of the physician when to say a patient has age-related macular degeneration.

Age-related macular deneration (AMD), however it be defined, is a major problem in our society. It accounts for a large percentage of visual disability among older individuals and is particularly disturbing because of its insidious onset and largely irreversible nature. This report will review the clinical characteristics of AMD and discuss the effects on the retina of light, an environmental factor that may account for some of these degenerative changes. We cannot reverse time, and we cannot shut our eyes completely to the light stimuli for which they exist but understanding the interaction between light exposure and the aging process may help us to minimize the damage in age-related eye disease.

THE CLINICAL PROBLEM: MACULAR DEGENERATION

Visual Loss in the Elderly

Visual loss is by no means inevitable among the elderly. Table 1 shows that 97% of individuals between 65 and 74 retain 20/40 or better acuity (nearly 92% are 20/25 or better). Even in the 75 to 85 age group, 87% have 20/40 vision or better. The dark side of this data is that a significant percentage (13% or more) of the population over 75 does not have sharp acuity, and the percentage rises with age. Furthermore, the statement that more than 90% of the 65-74 age group are 20/25 or better, hints at a more subtle problem: many of the elderly have good, but not quite perfect vision.

The Effects of Aging and Environment on Vision
Edited by D.A. Armstrong *et al.*, Plenum Press, New York, 1991

Table 1. DISTRIBUTION OF INDIVIDUALS BY AGE AND CORRECTED VISUAL ACUITY IN THE BETTER EYE (a).

Age Group	20/25 or better	20/40 or better
52 - 64	98.4 %	99.3 %
65 - 74	91.9 %	97.5 %
75 - 85	69.1 %	87.0 %

a. Data from the Framingham Eye Study (8).

As is discussed elsewhere in this book, many of the subtle aspects of visual perception such as contrast sensitivity, adjustment to light or dark, temporal resolution and form recognition are diminished with age. Thus, even an elderly individual who sees 20/20 under ideal conditions in a doctor's office may have some difficulty with practical tasks like recognizing faces at dusk or reading signs from a fast-moving car.

The clinical picture of AMD ranges from a subtle diminution of visual sensibility (that will probably affect all of us to one degree or another) to severe degrees of visual acuity loss that interferes with reading and driving, and causes functional and economic disability. For practical purposes, we speak of AMD as producing levels of acuity below 20/40 (the cut-off for driving tests), and acuity may fall to 20/400 or even to the point of barely counting fingers. Visual acuity of 20/200 (in the best corrected eye) generally qualifies as "legal blindness" and signifies a level at which reading is very difficult without a large magnifying glass or special aids. Despite this potential for severe visual loss, there is one redeeming feature to AMD: the damage is always limited to the macula. Peripheral vision invariably remains good, and even individuals with counting fingers vision can walk about comfortably and care for themselves. AMD never causes total blindness, and affected individuals can be firmly assured that they will always have some useful vision for getting around.

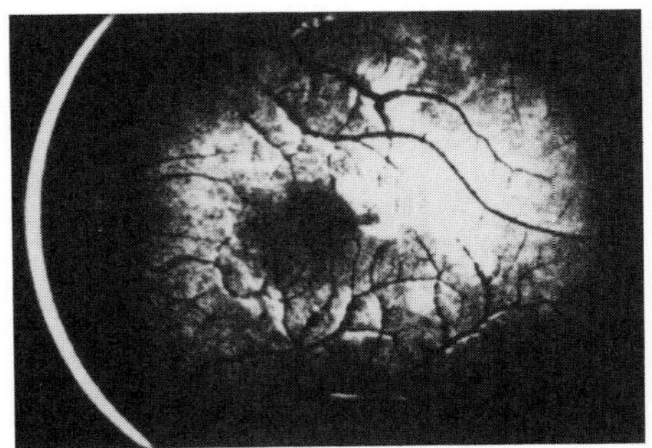

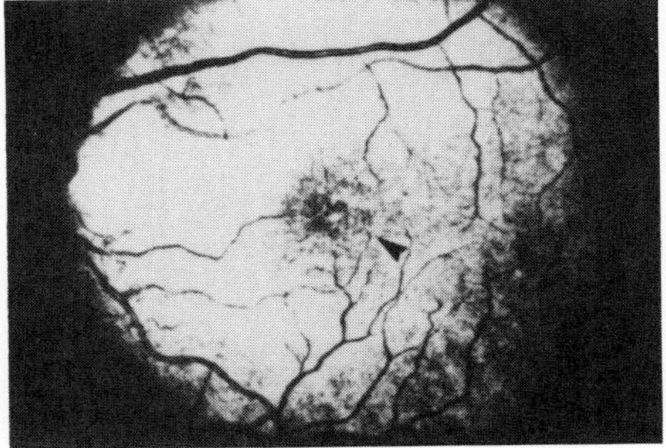

Figure 1. Changes in fundus appearance with age. Left: Retina in youth. Right: Retina in old age. Note loss of glistening reflexes, narrowed arteries and drusen (seen faintly near arrow).

Age related retinal damage

The retina and retinal pigment epithelium (RPE) show visible change with age (11): the ocular vessels appear narrower and more sclerotic than in youth, the RPE is less pigmented and allows the choroidal pattern to show through, and the youthful, glistening light reflexes in the macula and fovea are no longer seen (Figure 1). Yellow focal lesions called drusen are often present in the macula, and the peripheral retina may show a variety of degenerative changes such as reticular pigmentation, cobblestone atrophy and lattice degeneration. All of these anatomic alterations may be seen without any obvious loss of function, or various degrees of functional loss may be present without any more obvious ophthalmoscopic changes.

One reason for the difficulty in correlating visual function with anatomic changes is that the retinal neurons themselves, that determine perceptual ability, are clinically invisible. The retina is embryologically a part of the brain, and like brain tissue it loses cells with age. One recent study estimated that nearly 50% of the neurons on the visual cortex drop out by age 70 (1); although the loss rate for retinal neurons has not been accurately determined, retinal thickness decreases with age and typically about 20% of the photoreceptors are lost or damaged in the elderly eye. (5) This cellular loss may be an intrinsic property of retinal neurons or be secondary to vascular damage and other generalized facets of aging. Considering the amount of neuronal loss, it is not surprising that refinements in visual perception are diminished or lost with age, and it is remarkable that the elderly see as well as they do. (21)

The most serious loss of central acuity in AMD occurs when there is grossly visible anatomic damage. Sometimes the subtle process of atrophy extends to the point that large patches of RPE cells die leaving atrophic scars over which the photoreceptors have ceased to function (Figure 2). In other eyes the process of drusen formation continues to the point of pathologic breakdown in the RPE. Whereas isolated drusen are generally benign and carry little risk of visual loss, drusen that have become large and confluent (Figure 2) indicate a very significant risk of disastrous anatomic breakdown and scar formation (18). Confluent drusen may evolve into a detachment of the RPE, with fluid accumulating where it doesn't belong and the eventual formation of pigmented scars that signify permanent damage (6,7,16).

RPE detachment can lead to RPE cell damage that opens the normal barrier between the photoreceptors and their blood supply in the choroid. When this occurs, extracellular fluids, blood, and even new blood vessels themselves can cross the RPE and spread under the retina where they can cause irreversible damage. Exudative or hemorrhagic detachment of the retina is the most precipitous and serious form of macular degeneration because new vessels (called neovacularization) are abnormally fragile and even a small focus of them under the retina can cause massive bleeding and scarring. The diagnosis of subretinal neovascularization in AMD is best made with a test called fluorescein angiography in which the fluorescein dye is injected while the retina is photographed. The new vessels are abnormally leaky and neovascularization is readily identified (Figure 3). Once subretinal neovascularization has developed, central vision can be lost within only a few weeks time.

Therapy for Age-Related Macular Degeneration

Light protection may have a role in the long-range prevention of AMD, but once AMD has occurred therapy is limited to two approaches: the direct treatment of subretinal neovascularization (when it can be recognized early enough to prevent serious hemorrhage or exudation) and the provision of low vision aids (to maximize the use of whatever visual function remains).

A certain amount of misconception exists about laser treatment for AMD, because of publicity about the laser as a marvelous new tool in ophthalmology. Lasers have many uses, but in essence are no more than powerful light beams that burn or destroy tissue. Laser therapy is effective when a disease process can be localized to a small area of pathology, but lasers do not remove cataracts, stimulate new retinal cells to grow, or otherwise restore vision where retinal cells have been permanently damaged. If a focus of subretinal neovascularization can be identified early in the process of hemorrhagic AMD, before there has been bleeding under the fovea, then the laser can be used very effectively to destroy the new vessels and prevent further bleeding or scarring (10). This can be a sight-saving therapy, but it does not necessarily halt the underlying aging process or prevent subsequent degenerative changes.

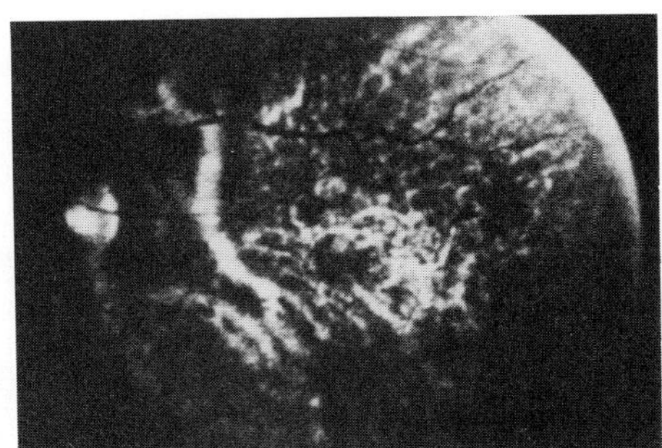

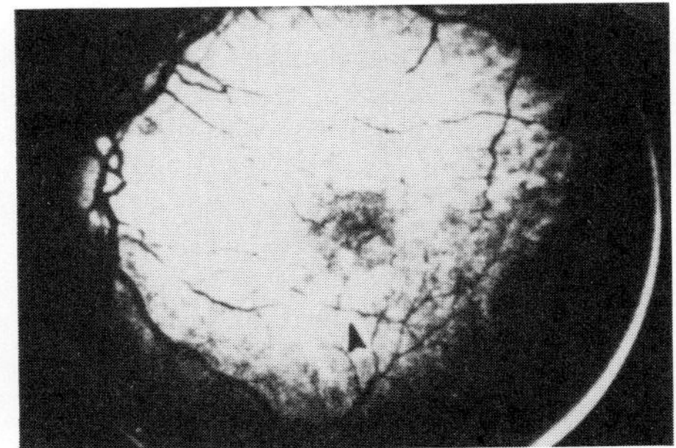

Figure 2. Age-related macular degeneration. Left: Dry, atrophic form (arrow). Right: Exudative form with large confluent drusen (arrow).

Realistically, most individuals with AMD have slow atrophic changes for with laser therapy is of no benefit. And many of the individuals with exudative or hemorrhagic AMD will, by the time their disease is recognized, either have damage that is too severe to treat or have new vessels growing right under the fovea where treatment would be as destructive as the disease itself.

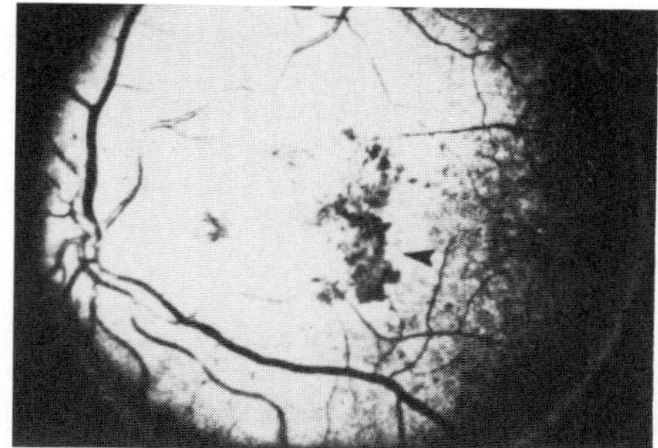
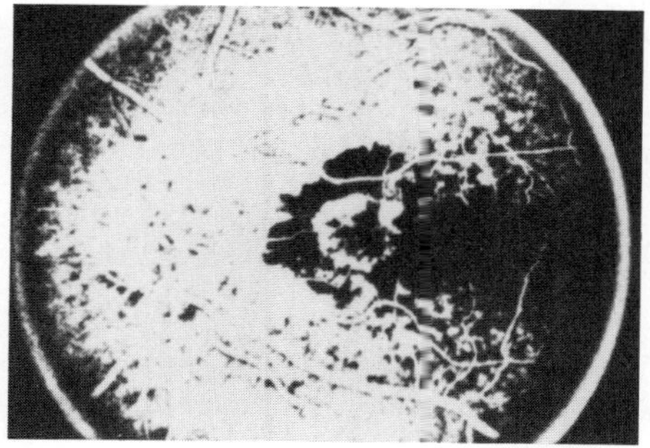

Figure 3. Hemorrhagic disciform macular degeneration with subretinal neovascularization. Left: Hemorrhage (arrow) extending under the fovea. Right: Fluorescein angiogram showing the network of new vessels (arrows) as a fuzzy ring (because dye leaks out of the fragile vessel walls).

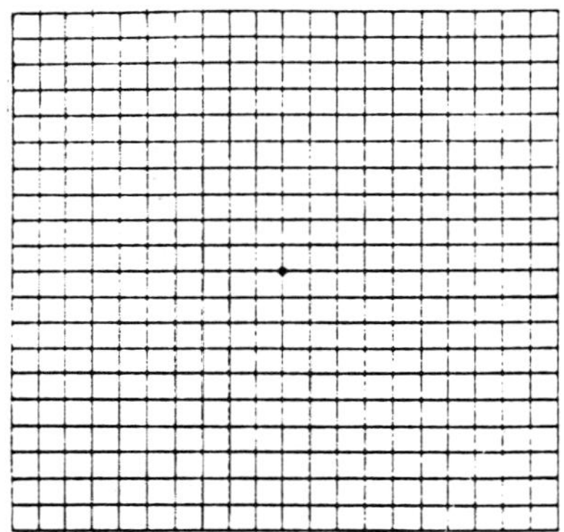
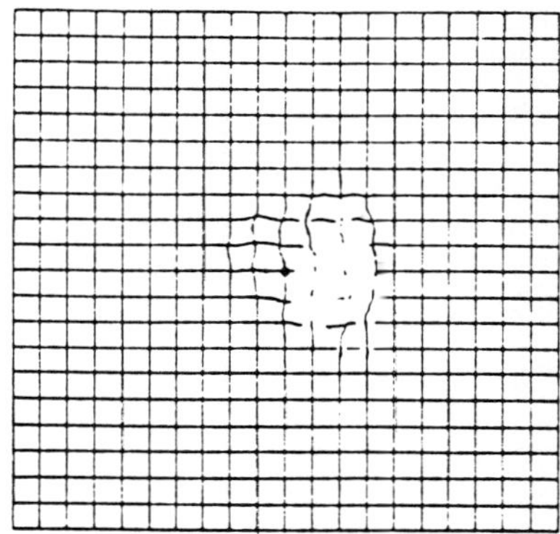

Figure 4. Amsler grid. Left: Normal appearance. Right: Grid as it may appear to someone with early macular degeneration.

The laser can actually be used in only about 10% of AMD cases -- but within this group therapy can be very effective. Thus, individuals with AMD should get in the habit of checking their own vision, in each eye independently, on a regular basis (e.g., every day or two), so that they can be aware of the earliest hint of visual loss. One way to do this is to pin up some newsprint or an Amsler grid (Figure 4) in the bathroom, and look at them as a part of the morning routine. Any definite change in acuity, or any new distortion (e.g. wavy lines) in the grid, is of concern and the individual should call an ophthalmologist immediately for evaluation and possible treatment. The most common laser used in the treatment of AMD uses argon gas to generate a green beam that is absorbed by RPE pigment and hemoglobin, and to some degree by the retina. Krypton lasers are also available now, and these generate red or yellow light that is absorbed much more by the RPE than by the overlying retina. This may allow more effective treatment of the neovascularization close to the fovea with less damage to retina that must be preserved for vision.

Whatever the cause of diminished acuity in AMD, a variety of visual aids are available that can improve visual function (2). Extra strong reading glasses are very effective for people with mild degrees of impairment. Hand magnifiers are useful for more severe degrees of loss, as are more elaborate devices such as closed circuit television cameras that can make reading possible for individuals with vision in the "count fingers" range. Further discussion of low vision management is beyond the scope of this paper, but suffice it to say that individuals with AMD should not be "written off" as blind or disabled, but should be carefully evaluated so that all opportunities for maintaining visual activity are explored.

LIGHT: A POSSIBLE PATHOGENETIC FACTOR

Light Energy: Light (Figure 5) is electromagnetic energy, comprising the ultraviolet (wavelength below 400 nanometers), the visible spectrum (400 to 700 nm), and the infrared (greater than 700 nm). For light to be perceived by an eye, or indeed for light to be absorbed in any medium, the energy must be transduced (i.e., changed) to a different form. This may represent a transformation to heat, evident in the temperature of a black object out in the noonday sun, or dissipation of energy through chemical and physical molecular reactions.

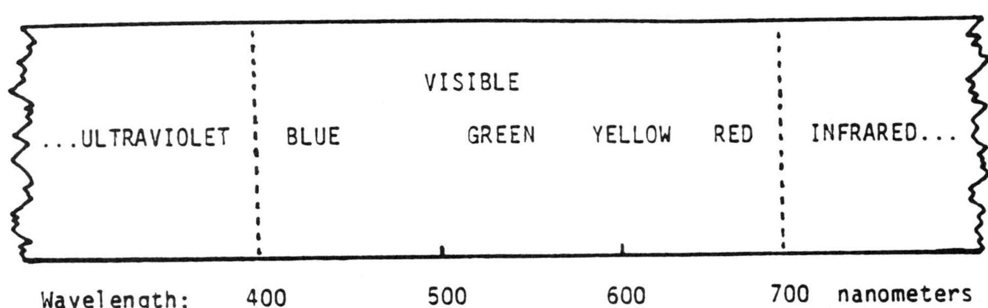

Figure 5. Light as a part of the electromagnetic spectrum.

Visible light is absorbed by special visual pigments in the photoreceptor cells, and the chemical change induced in these pigmented initiates the nerve signal which ultimately reaches the brain. Each of the different visual pigments in the eye absorbs certain wavelengths more readily than others, but once a photon of light (the elemental unit of light energy) is absorbed, its chemical effect is the same regardless of whether the light is red, green, or blue. The situation may be different, however, when light is absorbed by other proteins and lipids for which the degree of light-induced may be proportional to the energy of each photon. Short wavelengths have the highest energy, so that the blue or ultraviolet light is generally more damaging than light at the red end of the spectrum.

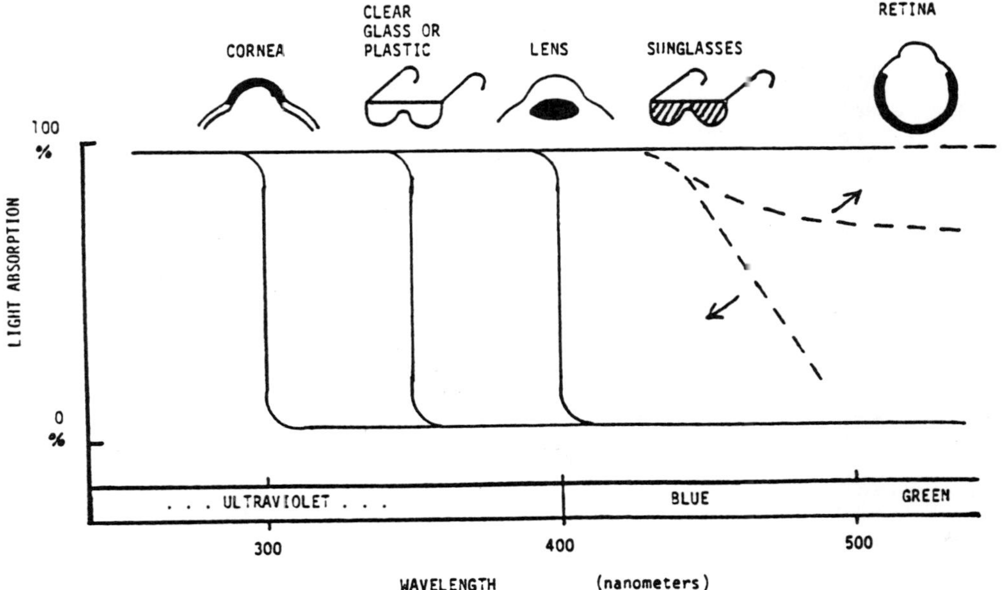

Figure 6. Absorption of light in the eye: The cornea, clear glasses and the lens block out progressively longer ultraviolet wavelengths. Good quality sunglasses should block all of the ultraviolet, as shown in the diagram.

Although all wavelengths of light impinge upon the cornea, the degree of penetration into the eye (14) is highly dependent upon wavelength (Figure 6). The cornea absorbs ultraviolet wavelengths up to 300 nanometers in length, which explains why unshielded sun exposure may produce a painful corneal burn, much like sunburn on the skin.

Spectacles or contact lenses, made out of either glass or plastic, absorb ultraviolet up to about 350 nanometers. Of the light which passes through the cornea, the human lens absorbs wavelengths up to 400 nanometers, which includes the deep blue end of the visible spectrum. Thus, for practical purposes, the retina receives mostly light above 400 nanometers. However, if an individual is aphakic (i.e., has had the lens removed), near ultraviolet and deep blue light will also reach the retina. Many of the infrared wavelengths also pass through cornea and lens, but the energy levels from ordinary sources are not sufficiently high to generate heat in ocular tissues.

The photoreceptor renewal cycle

How do body tissues survive a lifetime of radiant energy exposure? Clearly no self-respecting piece of skin would try to last for 70 years under the assault of countless hours of sun, not to mention wind, skin lotions and abrasion. Our skin survives by growing new cells continually while the old ones rub off and are discarded. The retina in the back of the eye is also exposed to a lifetime of light, but obviously cannot slough its cells to the outside. How, then, does retina survive and regenerate itself?

The answer lies in a cycle of synthesis and digestion involving the photoreceptors and the RPE (22). From the standpoint of potential light damage, the most critical components of the retina are the photoreceptors, the primary cellular elements which receive light and change it to a neural signal. These cells consist of an <u>inner segment</u> that synthesizes molecular materials, and an <u>outer segment</u> composed of densely packed lipid membranes in which the visual pigment molecules are embedded. The photoreceptors lie adjacent to the RPE which separates them from their primary blood supply in the choriocapillaris and regulates their nutritional environment. To renew themselves, the inner segments continually synthesize new outer segment membranes, complete with the requisite visual pigments. Concomitantly, outdated membranes at the tips of the outer segments are absorbed at regular intervals into the adjacent RPE where they are digested by cellular enzymes (Figure 7). These enzymes break down the photoreceptor materials into components that can be excreted into the choriocapillaris or recycled back to the photoreceptor inner segments to be reincorporated into new photoreceptor membranes.

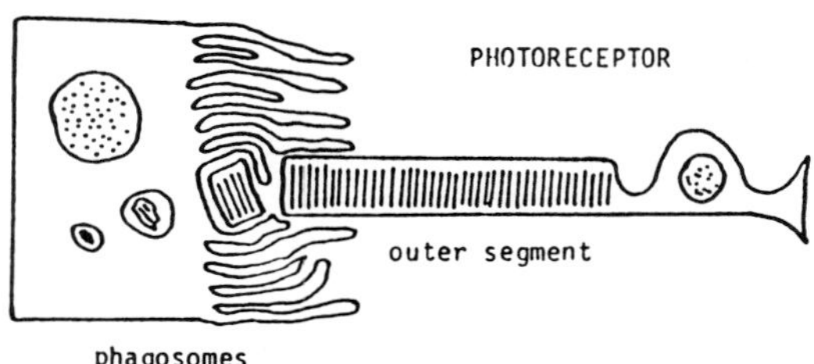

Figure 7. Phagocytosis of outer segment tips by the retinal piment epithelium. The encapsulated vesicles of outer segment material are called phagosomes, and they become progressively smaller as lysosomal enzymes digest the outer segment material.

Aging in a stressful environment

We noted earlier that physical changes in the RPE such as a decrease in the degree of melanin pigmentation, are visible in virtually all older eyes. There is also the gradual accumulation of an "aging pigment," called lipofuscin that first appears in the RPE in youth, and by the middle age or beyond has clogged the RPE cells with brownish-yellow, autofluorescent material (4). Its mere presence has no proven detrimental effect, but it probably signifies a degree of failure in the outer segment renewal cycle and thus may relate to more serious degenerative changes within the RPE. Lipofuscin is chemically related to the lipid structure of outer segment membranes, and is thought to represent poorly digested lipids from the outer segment renewal cycle. As the amount of cytoplasmic waste material becomes oppressive, some RPE cells attempt to eject it towards the choriocapillaris with a resulting accumulation of cellular debris immediately beneath the RPE. Large excrescences of debris (Figure 8) are visible clinically as drusen.

One of the factors which is thought to contribute to membrane damage, and inability of the RPE to fully digest membrane debris, is oxidation. (3) Body tissues in a high oxygen environment tend to form activated oxygen molecules such as singlet states and oxygen free-radicals These molecules are highly reactive, and produce lipid peroxides and lipid free-radicals that, in the manner of a chain reaction, react with other lipids to create new reactive states. The photoreceptors are particularly at risk for this type of oxidative chain reaction because they have an unusually high concentration of lipid material in the outer segments, and they lie in an unusually high oxygen environment. The choriocapillaris has the highest blood flow per unit are of any tissue in the body and bathes RPE and photoreceptors in an excess of oxygen so that there is always an adequate supply even under conditions of low blood pressure or slowed breathing.

Fortunately, a variety of bodily defense mechanisms exist to absorb or break down reactive oxidative molecules. These include anti-oxidant substances such as Vitamin E, Vitamin C, catalase (a zinc-requiring enzyme) and glutathione peroxidase (a selenium-requiring enzyme). Under most conditions, these agents keep oxidative damage to a minimum, or at least to a level at which normally bodily defense mechanisms can cope with damaged material and resynthesize necessary substances. However, the cumulative effects of minimal degrees of such damage occurring over a lifetime of oxidative stress may well be one component of aging.

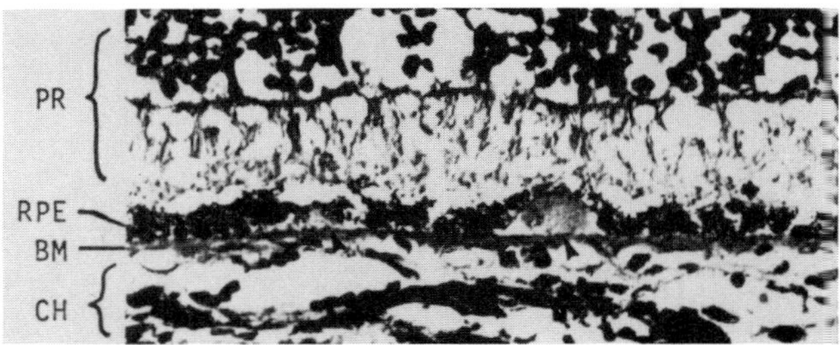

Figure 8. Light micrograph of drusen (arrows), visible as lumps of hyaline material under thinned retinal pigment epithelial (RPE) cells. (PR = photo-receptors; BM = Bruch's membrane; CH = choroid).

Light contributes to this equation because the absorption of radiant energy enhances the formation of oxygen-free radicals. Thus, retina is at risk not only from concentrated lipids and high oxygen, but exposure to light. The true significance of light in aging, however, is unknown. We do not know whether ordinary environmental light stress is benign because antioxidants and recovery processes within the retina compensate for it completely, whether ordinary light exposure contributes to a slow accumulation of waste products that damage the RPE over a lifetime, or whether the greatest risk is from occasional brief exposures to high levels of light on the beach or by the swimming pool. We can merely recognize that light contributes to the process of oxidative damage which we think contributes to a formation of lipofuscin and other concomitants of aging in the eye.

Light damage

Although the role of light in aging remains speculative (however reasonable it may seem) there is unequivocal evidence that strong or prolonged light exposure is damaging to animal eyes (9,14). Broadly speaking, two types of damage have been described: 1) Visible light damage that primarily affects the photoreceptors, and 2) Short wave and blue light damage which affects both photoreceptor and RPE.

Visible light damage was first described by Werner Noell(15) who observed that albino rats, kept for a day or more in moderately bright fluorescent lighting, showed degeneration of their photoreceptors. Albino animals seem peculiarly susceptible to this type of damage, presumably because they lack the screening protection of melanin in their RPE and choroid. This type of damage is produced most effectively with blue-green light, which is the color absorbed most readily by the rod visual pigment, rhodopsin. Visible light damage is rather difficult to produce in pigmented animals, and requires quite prolonged and stressful light exposures. Thus, its role in human retinopathy or aging is unclear. It maybe of concern when individuals are exposed all day to bright sunlight without protection, as may occur if one neglects to wear sunglasses on the beach or in the mountains.

Short wavelengths and blue light damage is thought to be more relevant to aging (14, 20), since it damages not only photoreceptors but the RPE, and since the light may affect proteins and lipid membrane structures other than the visual pigments. Animals exposed to intense bright lights will develop immediate vacuolization and cellular disorganization in both photoreceptors and RPE, and the damage is most intense with shorter wavelengths. The injury is "photic" rather than thermal, meaning that it occurs at levels of energy far below those necessary to elevate tissue temperature and produce a burn (as takes place during laser photocoagulation). If retinal or RPE tissue is exposed in vitro, the most severe photic damage is caused by ultraviolet wavelengths, but the situation is different in the living eye because the cornea and lens filter out most of the ultra-violet. In the intact eye, the peak wavelength for short wavelength damage is 435 nm, which is well within the visible blue end of the spectrum (17). In an aphakic eye, which lacks the lens and its filtering power, the peak wavelength for damage is below 400 nm and thus lies in the ultraviolet range. Short wavelength light is believed to cause its damage by enhancing free radical formation and contributing to the cycle of oxidative damage within the eye. Consistent with this view, the effects of intense light exposure can be potentiated by making experimental animals deficient in antioxidants such as vitamin E or antioxidant enzyme cofactors such as zinc and selenium (19).

Under ordinary conditions, even outdoors, most people do not encounter damaging levels. However, prolonged unprotected exposure to diffuse sunlight (e.g. on snow or water), moderately long exposure to light focused on the retina (e.g. during eye examination or surgery), or even brief staring at the sun can bring an eye into the range of potential damage. For this reason, cataract surgeons use ultraviolet filters in the operating microscope and cover the cornea with a protective shield during much of the procedure. I should emphasize that no damage should occur from ordinary ophthalmic examinations, but if diagnostic instruments were left focused upon one place in the retina for several minutes or longer, the damage threshold might be approached.

If light really contributes to aging, one would think that individuals living outdoors in sunny areas of the world would have a much higher incidence of macular degeneration that those working indoors in the Northern Hemisphere. However, no convincing study on this hypothesis has been published, perhaps because this type of epidemiologic study is extremely difficult to construct and control. Unless the subject populations are genetically identical, and additional factors such as diet, use of sunglasses, etc., are known, the findings may bear little relationship to light exposure.

Finally, lighting conditions may affect visual function in ways other than direct damage to the photoreceptors in RPE (13). Aging eyes, for example, are much more susceptible to visual interference from glare and bright backlighting because of increased cloudiness of the cornea and lens. Older eyes show slower and less complete adaptation to light and dark, and will recover more slowly to any sudden or dazzling stimuli. Some of these symptoms may be subjectively, if not objectively, dependent on wavelength. For example, bluish light, such as from sunlight or fluorescent lighting is generally more uncomfortable and glare-producing than warm-toned incandescent lighting. Whereas some of these difficulties undoubtedly derive from haziness in the media, which scatters light, some may derive from the gradual loss of neurons and RPE cells with age that impairs the more complex and subtle integrative aspects of visual perception.

Light protection and environmental design

At the cellular level, light damage--and possibly part of the intrinsic aging process--involves oxygen free radical formation and oxidative damage. Theoretically, if one could inhibit these oxidative reactions or at least quench their activity, part of the aging process could be prevented. Experimental work has shown clearly that deficiency in antioxidant substances puts tissue at great risk for both light and oxidative damage, and thus we can extrapolate that clinical deficiency in Vitamin E, Vitamin C, zinc and selenium should be avoided. The average American eating a good diet should not be deficient in any of these substances, however, and ordinary vitamin supplementation would be in any event be more than adequate to insure reasonable levels. Unfortunately for our dreams of everlasting youth, many studies have shown that higher doses of these substances, beyond normal levels, do not confer any additional light protection or antioxidant benefit. Furthermore, megavitamin therapy can be hazardous because high levels of Vitamin E, zinc and selenium are toxic.

Insofar as light absorption, particularly at the blue end of the spectrum, contributes to oxidative damage, we can achieve a measure of protection through the use of sun visors and appropriate filtering lenses. The simple expedient of wearing a cap on a sunny day eliminates a large amount of blue and ultraviolet radiation from the sky. It is probably advisable for everyone to use good quality sunglasses under bright outdoor conditions, i.e. sunglasses that specifically block the full UV spectrum and the deep blue visible spectrum (Figure 9). Unfortunately, it is not always easy to determine whether sunglasses accomplish this, since there are no standards or either manufacture or labeling (20). All glass and plastic lenses (even transparent ones) block the far UV (i.e. below 350 nm) and some manufactures have used this fact to claim that their sunglasses are "UV-absorbing." Careful inquiry may be necessary to judge the quality of a lens. In general optical quality glasses are more effective than cheap, drug-store varieties; also, many UV-absorbing lenses will have a yellowish or brownish cast because they absorb at the blue end of the visible spectrum.

The person who has had cataract surgery is in a special situation, since the normal UV-absorption of the lens has been lost. Theoretically such eyes are at greater risk to suffer light damage and accelerated aging, and it seems reasonable to use UV absorbers (which are available) in spectacles, contact lenses or implanted intraocular lenses for such eyes. UV-absorbing plastic can also be used as a base material for sunglasses to insure that full UV protection is obtained. These expedients may be of greater benefit in restoring normal color balance and reducing glare than in protecting the retina, however, in a population that is already aged and spends little time unprotected in bright sunlight. There has never been any epidemiologic evidence that aphakes have more macular degeneration that phakic individuals, and by the time most cataract surgery is done the horse is out of the barn with respect to aging changes. We would be kidding ourselves to think that providing light protection to a 70 year old will magically stop the degenerative process. If macular changes are to be prevented and light plays any role, a lifetime of protection will probably be required. Nevertheless, it is still rational to provide ultraviolet protection to aphakes--and even more important to provide it to individuals that have retinal dysfunction at a younger age from intrinsic retinal disease such as juvenile macular dystrophy or retinitis pigmentosa. (12)

Since outdoor lighting intensities are usually far below those outdoors, short wavelength light damage is not ordinarily a major concern in designing environmental lighting except in industrial situations where intense light sources exist. There are other ordinary environmental lighting concerns, however, which are very relevant to vision. Older individuals may have great difficulty seeing under conditions that are still acceptable to the young, because of the scattering of light in the media and the loss of fine tuning in visual perception.
Older individuals will have difficulty going from one level of lighting to another, and may complain of glare or interference from focal lighting sources. Paradoxically, they may at the same time require more light than others to read or see clearly. One often finds conflict between needing more illumination to see, yet less to avoid glare and dazzle. Environmental lighting design for areas that may be used by older individuals--or indeed any individual with retinal disease or cloudy media--should try as much as possible to resolve this conflict: Ambient light intensities should be high

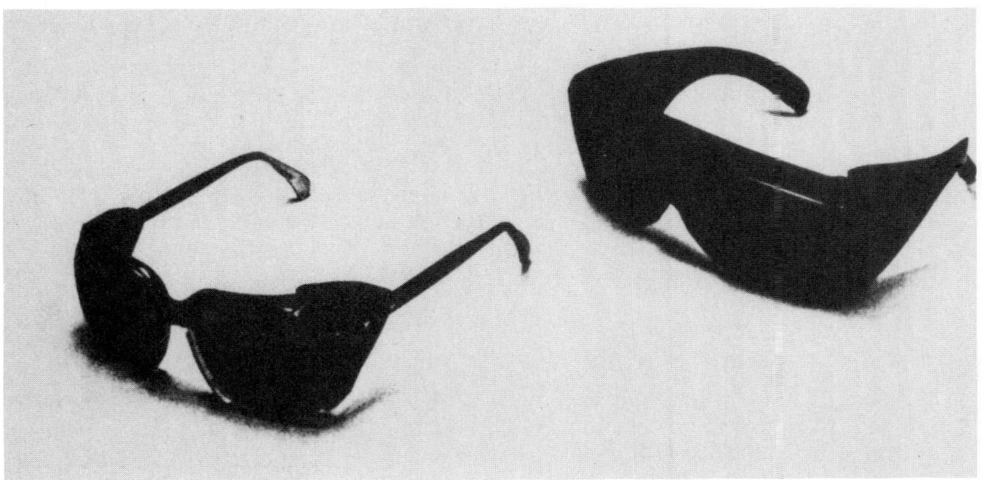

Figure 9. Examples of UV-absorbing glasses with side and top shields. Illustrated are orange Corning 550 lenses(left) and amber Solarshield goggles(right), but a number of brands are available.

enough to facilitate reading, but the light sources should be biased toward the warm tones and should be diffuse to avoid reflection off signs or reading surfaces. Shadowy areas should be avoided so that people do not have to see in dim illumination immediately after being adapted to higher light intensities. Reasonable care to these considerations will make the environment much more habitable to individuals with visual dysfunction.

CONCLUSIONS

Age related macular degeneration (AMD) causes significant disability in our older population. In many individuals this is an indolent atrophic process; in others visual loss occurs from exudation and hemorrhage that results from the growth of abnormal new vessels through a damaged retinal pigment epithelium. If detected early, this hemorrhagic type of AMD can sometimes be treated by laser photocoagulation. Even though significant visual loss may occur in AMD, functional vision may be possible with visual aids. Fortunately, AMD never extends beyond the macula (and patients will always retain good side version for getting around).

The inevitability of aging may be modified by factors such as tissue oxidation and light-exposure. The role of light has not been proven with certainty, but circumstantial evidence suggests that it contributes over a lifetime to degenerative processes in the retina. Protection from bright exposures (particularly of short wavelength light) is prudent for all of us and is medically indicated for people with aphakia or retinal disease. Well-designed environmental lighting may not affect aging but can improve visual function for the elderly.

REFERENCES

1. Devaney, K.O., and Johnson, H.A., 1980, Neuron loss in the aging visual cortex, J. Gerontol., 15:836.
2. Faye, E.E., 1976, "Clinical Low Vision," Little, Brown & Co., Inc., Boston.
3. Feeney, L., and Berman, E.R., 1976, Oxygen toxicity: Membrane damage by free radicals. Invest. Ophthalmol., 15:789.
4. Feeney-Burns,L., Berman, E.R., and Rothman, H., 1980, Lipofuscin of human retinal pigment epithelium, Am. J. Ophthalmol., 90:783.
5. Gartner, S., and Henkind, P., 1981, Aging and degeneration of the human macula. 1. Outer nuclear layer and photoreceptors, Br. J. Ophthalmol., 65:23.
6. Gass, J.D.M., 1973, Drusen and disciform macular detachment and degeneration, Arch. Ophthalmol., 90:206.
7. Grindle, C.F.J., and Marshall, J., 1987, Aging changes in Bruch's membrane and their functional implications, Trans. Ophthalmol. Soc. UK, 98:172.
8. Kahn, H.A.,Leibowitz, H.M., Ganley, J.P., Kini, M.M., Colton, T.,Nickerson, R.S., and Dawber, T.R., 1977, the Framingham Eye Study. 1. Outline and major prevalence findings, Am.J. Epidemiol., 106:17.
9. Lanum, J., 1978, The damaging effects of light on the retina. Emperical findings, theoretical and practical implications, Surv. Ophthalmol., 22:221.
10. Macular Photocoagulation Study Group, 1982, Argon laser photocoagulation for senile macular deneration, Arch. Ophthalmol., 100:912.
11. Marmor, M.F., 1982, Aging and the retina, in: "Aging and human visual function," Sekuler, R., Kline, D., and Dismukes, K., eds., Alan R., Liss, New York.
12. Marmor, M.F., et al, 1983, Retinitis pigmentosa. A Symposium on terminology and methods of examination, Ophthalmol., 90:126.

13 Marmor, M.F., 1986, Visual changes with age, in: "The eye and its disorders in the elderly," Caird, F.I. and Williamson J., eds., John Wright & Sons, Bristol.
14. Marshall, J., Radiation and the aging eye. 1985, Ophthal. Physiol. Opt., 5:241.
15. Noell, W.K., Walker, V.S., Kang, B.S., and Berman, S., 1966, Retinal damage by light in rats, Invest. Ophthalmol. Vis. Sci., 5:450.
16. Sarks, S.H., 1976, Aging and degeneration in the macular region: A clinico-pathological study., Br. J. Ophthalmol., 60:324.
17. Sliney, D.H., and Wolbarsht,M.L., 1980, Safety standards and measurement techniques for high intensity light sources, Vision Res., 20:1133.
18. Smiddy,W.E., and Fine, S.L., 1984, Prognosis of patients with bilateral macular drusen, Ophthalmol., 91:271.
19. Stone, W.L., Katz, M.L., Lurie, M., Marmor, M.F., and Dratz, E.A., 1979, Effects of vitamin E and selenium on light damage to the rat retina, Photochemistry and Photo-biology, 29:725.
20. Waxler, M., and Hitchens, V.M., eds., 1986, Optical radiation and visual health, CRC Press, Boca Raton.
21 Weale, R.A., 1982, "A Biography of the Eye: Development, Growth, and Age," H.K. Lewis, London.
22. Young, R.W., 1976, Visual cells and the concept of renewal, Invest. Ophthalmol. Vis. Sci., 15:700.

BIBLIOGRAPHY

1. Gawande, A.A., Roloff, L.W., and Marmor, M.F., 1991, The specificity of colored lenses as visual aids in retinal disease, J. Vis. Impairment and Blindness (in press).

2. Mainster, M.A., 1987, Light and macular degeneration: A biophysical and clinical perspective, Eye, 1:304.

3. Marmor, M.F., 1990, New directions in the management of ocular disease, in "Opthalmology Clinics of North America", Masse, S., ed., W.B. Saunders, Co., Philadelphia.

4. Munoz, B., West, S., Bressler, N., Bressler, S., Rosenthal, F.S., and Taylor, R., 1990, Blue light and risk of age-related macular degeneration. Invest.Opthalmol. Vis. Sci. (Supp.), 31:49.

5. Newsome, D.A., Swartz, M., and Leone, N.C., Elston, R.C., and Miller, E., 1988, Oral zinc in macular degeneration. Arch. Opthalmol., 106:192.

6. Tso, M.O.M., 1989, Experiments on visual cells by nature and man: In search of treatment for photoreceptor degeneration (Friederwald lecture). Invest. Opthalmol. Vis. Sci., 30:2430.

7. West, S.K., Rosenthal, F.S., Bressler, N.M., Bressler, S.B., Munoz, B., Fine, S.L., and Taylor, H.R., 1989, Exposure to sunlight and other risk factors for age-related macular degeneration, Arch. Opthalmol., 107:875.

8. Yoon, Y.H., and Marmor, M.F., 1989, Dextromethorphan protects retina against ischemic injury in vivo, Arch. Opthalmol., 107:409.

9. Young, R.W., 1988, Solar radiation and age-related macular degeneration, Surv. Opthalmol., 32:252.

UPDATE 1990

Debate has continued on the role of light as a factor in the pathogenesis of macular degeneration. Biophysical and animal experimentation data argues that light contributes to macular aging [2,9]. A carefully controlled study of Chesapeake Bay watermen showed no relationship between cumulative UV-A or UV-B exposure and the development of macular degeneration [7] but men with severe disciform degeneration had experienced more blue light exposure [4].

There is still no definitive labelling standard for sunglasses, and UV-absorbing properties vary widely with no relationship to cost. The Sunglass Manufacturers of America have labelled some products as "General Purpose" or "Special Purpose" sunglasses, but both ratings may allow passage of up to 40% of the potentially damaging UV-A rays. The same concern applies to lenses which "meet the ANSI Z-80 Standard". The best policy is still to buy a product which explicitly excludes all UV, including UV-A, or all radiation up to a desired wavelength (e.g. 400 rm to exclude all UV, 450 rm to exclude deep blue).

Glasses which block blue may give subjective comfort to some patients (just as yellow goggles may seem tohelp vision on the snow). Light orange or yellowish filters seem most effective for individuals with macular degeneration, but no one color is ideal for all patients[1]. A trial of colored lenses under real-life conditions is necessary to determine whether any are of value to the individual. There is continued interest in oxidative damage, with or without light injury, as an adjunct to macular aging. High levels of vitamin C can protect against experimental light damage [6]. One report [5] suggested that high-dose zinc therapy improved visual function in patients with age-related macular degeneration, even though serum zinc levels were normal. However, visual benefits from vitamin C therapy have not been demonstrated in man, and the zinc report is puzzling and has yet to be verified. Most specialists remain cautious about pharmacologic therapy for macular degeneration and recommend at most low-dose vitamins to avoid deficiency. Antioxidant vitamin-mineral products are now targeting the eye market, and one should be wary of therapeutic claims or implications.

The future will yield new information on the pathophysiology and clinical evaluation of macular degeneration [3]. For example, better strategies for coping with aging and environmental damage to cells may come out of knowledge about how cells respond to injury and why they die. Temporary ischemia can kill retinal cells as a result of secondary release of toxic neurotransmitters, but this cell death may be prevented by blocking the transmitter receptors[8].

SENESCENT ALTERATIONS IN THE RETINA AND RETINAL PIGMENT EPITHELIUM: EVIDENCE FOR MECHANISMS BASED ON NUTRITIONAL STUDIES

Martin L. Katz[1] and W. Gerald Robison, Jr.[2]

[1]University of Missouri School of Medicine
Mason Institute of Ophthalmology
Columbia, Missouri 65212

[2]National Eye Institute
National Institutes of Health
Bethesda, Maryland 20892

INTRODUCTION

Age-related diseases of the retina are among the most prevalent causes of serious visual impairment in developed countries. Those retinal diseases affecting the macular region have the most pronounced effects on visual capacity. Data from an epidemiological study conducted by the National Center for Health Statistics indicate that the incidence of age-related macular degeneration (AMD) in the United States is approximately 85 per 1,000 of the population 65 to 74 years of age (11). In the Framingham Eye Study, the incidence of AMD was found to be 110 per 1,000 of population in the same age group, and rose to 280 per 1,000 in those 75 to 85 years of age (34). At present, our understanding of the etiology of AMD is extremely limited, but it is likely that age-related diseases of the macula develop as the result of underlying senescent alterations in the retina and associated tissues. If we are to retard or prevent the development of AMD, therefore, it will be necessary to identify primary senescent changes that may lead to the development of macular degeneration.

It is possible that AMD develops directly as a result of primary age-related changes in the structure and function of retinal neurons. Senescent alterations in tissues serving support functions for the retina may also be involved in the development of AMD. One of the most important retinal support tissues is the retinal pigment epithelium (RPE). The RPE consists of a single sheet of epithelial cells which lies between the photoreceptor cells in the nueral retina and their supporting blood supply in the choriocapillaris (Fig. 1). Impairment of RPE cell function is known to result in photoreceptor cell degeneration (40). Therefore, age-related changes in RPE function could have detrimental effects on the neural retina. Retinal and choroidal blood vessels also play critical roles in maintaining

The Effects of Aging and Environment on Vision
Edited by D.A. Armstrong *et al.*, Plenum Press, New York, 1991

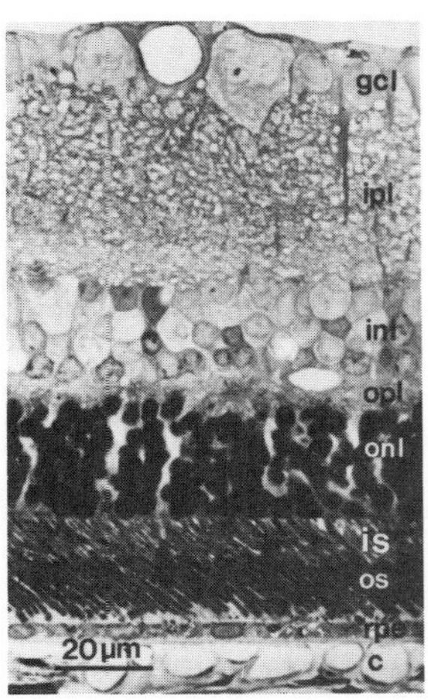

Fig. 1. Light micrograph of a rat retina in cross-section. The retina is organized into a number of distinct layers, which include the ganglion cell layer (gcl), inner plexiform layer (ipl), inner nuclear layer (inl), outer plexiform layer (opl), and the photoreceptor cell layer consisting of the photoreceptor cell nuclei in the outer nuclear layer (onl), and the inner segment (is) and outer segment (os) portions of the photoreceptor cells. The retinal pigment epithelium (rpe) forms a continuous sheet between the photoreceptor cells and the choriocapillaris (c).

retinal metabolism; therefore, age-related alterations in the retinal vasculature would also be expected to affect retinal function. Because age-related alterations in either the neural retina or its supporting tissues may underlie AMD, studies have been carried out to identify changes which occur in these tissues during senescence, and to determine potential mechanisms underlying these changes. Particular interest has been focused recently on possible relationships between nutrition and age-related alterations in the retina and associated tissues.

AGE-RELATED CHANGES IN THE RETINA, CHOROID, AND RPE

A gradual loss of retinal neurons appears to accompany the aging process in humans and other mammalian species. Based on determinations of axon densitities in the optic nerve, Balazsi and colleagues (2) reported that over a 70-year lifespan, approximately 25% of the ganglion cells are lost from the human retina. Photoreceptor cell densities in the human retina have also been reported to decline during senescence (12,36). It is likely that these cell losses contribute to the development of age-related declines in visual function, such as the decreased light sensitivity reported by Haas and colleagues (14) and the decline in visual acuity reported by Jay et al. (20). Age-related decreases in retinal neuron density, similar to those reported in humans, have been observed in rhesus monkeys (45) and in a number of rat strains (29,38,50,52). Therefore these species may be useful models for studying the mechanisms underlying the loss of neurons from the human retina during senescence.

Most mammalian retinas, including those of the human and rat, are served by two separate capillary networks. Within the neural retina itself is a bed of capillaries which provides support mainly for retinal neurons post-synaptic to the photoreceptor cells (3). A large bed of capillaries just exterior to the retina, on the other hand, is the major source of nutrients and waste removal for the photoreceptor cells (3). Age-related alterations in perfusion of these capillaries, or in exchange of materials between the lumens of the capillaries and the tissues they serve, could be partially responsible for the observed losses of retinal neurons. A variety of evidence suggests that aging may be accompanied by a deterioration in the vascular support system of the retina.

Capillaries within the neural retina have been reported to undergo an age-related loss of both endothelial cells and pericytes in human (37) and rat retinas (13). As a result of this cell loss, some capillary regions become totally acellular, and are apparently occluded. Thus, the perfusion of the neural retina appears to become less extensive as a result of aging. Retinal capillary basement membranes also increase in thickness during senescence, and this thickening may impair the exchange of materials between the capillary lumens and the surrounding tissues (36,44). The age-related alterations in the retinal capillaries may be involved to some degree in the observed loss of neurons from the inner layers of the retina during aging.

Age-related changes in the choroidal vasculature have been less well characterized than those of the retinal vessels. However, extensive studies have been carried out on age-related changes in Bruch's membrane and the RPE, which are located between the choroidal vessels and the retinal photoreceptor cells that are supported by the choroidal vasculature. Bruch's membrane is composed of several layers of connective tissue sandwiched between the basal plasma membrane of the RPE and the endothelial cells of the choroidal capillaries. The connective tissue components of Bruch's membrane undergo a number of morphological alterations as a result of the aging process. The short collagen fibers and elastic filaments within Bruch's membrane become coated with an electron-dense substance in the human eye during senescence (17,36) In addition, particles and vesicles of varying appearance are deposited within Bruch's membrane (19,36). Localized thickening of the central layers of Bruch's membrane also occurs during aging in the human eye (18,35,36); these thickened areas are called drusen. RPE cells overlying drusen have been reported to undergo degenerative changes in the human eye (10). The basal laminae of the RPE and the choroidal endothelial cells, which form the outer boundaries of Bruch's membrane, also become progressively thicker during the aging process (27,36), and the overall thickness of Bruch's membrane increases substantially as well (19). Although the significance of these age-related alterations in Bruch's membrane for retinal function are currently unknown, it is possible that these changes may result in an impairment of the exchange of materials between the retina and the choroidal vessels.

Because the RPE plays a number of significant roles in retinal metabolism, a great deal of effort has been focused on characterizing age-related alterations in RPE cell structure and function, and in elucidating the mechanisms underlying these changes. One of the most prominent changes undergone by the human RPE during senescence is the progressive intracellular accumulation of the autofluorescent pigment, lipofuscin (6,8,51,53,54). A similar pigment accumulates during aging in a wide variety of post-mitotic cell types in every animal that has been examined. This observation suggests that lipofuscin deposition and the aging process are associated in a fundamental manner. Other age-related changes that have been reported to occur in the RPE of the human eye include a decrease in melanin content (7,8,53), and an increase in lipid-body content (18). Both enlargement and flattening of the RPE basal infoldings have been reported to occur in human eyes with advancing age (18,43). Within the basal infoldings, there is an age-related deposition of material that is continuous with, and similar in appearance ot the RPE basal lamina (18). Banded fibers with a unique and distinctive appearance are also deposited within the basal infoldings of the human RPE (18, 49). Accompanying these changes in RPE morphology is the development of an irregularity in cell size and shape with advancing age (10,43).

Many of the morphological alterations undergone by the human RPE during senescence have also been observed to occur in the rat (27). An essentially linear relationship between RPE lipofuscin content and age has been reported in pigmented rats (27) (Fig. 2). The lipid-body content of the rat RPE also increases during senescence (27). As in the human, there are age-related alterations in the morphology of the RPE basal infoldings in pigmented rats (28). The RPE basal lamina thickens dramatically during the aging process, and often extends up into the basal infoldings (27). Banded fibers, similar to those observed in human eyes, are deposited within the basal infoldings (27). RPE cell height in the central retina increases significantly during aging in the rat (27). The apical microvilli of the RPE undergo a substantial morphological alteration during aging, changing from flattened, sheet-like structures in young rats, to thicker, tubular extensions in senescent animals (27). The latter age-related alteration in RPE morphology has not been reported to occur in human eyes.

In addition to morphological alterations, the rat RPE also appears to undergo a number of functional changes during the aging process. As part of normal renewal, photoreceptor outer segment discs, which contain the photosenstitive pigment responsible for vision, are continually being added to the bases of the outer segments (Fig. 3). Groups of the oldest disc membranes are periodically shed from the apical ends of the outer segments and are then engulfed and degraded by cells of the RPE. Shedding of rod photoreceptor disc membranes follows a circadian pattern, with the peak occurring within 1 to 2 hours after light onset in rats maintained in a cyclic lighting environment (39,41). Impairment of this outer segment renewal process has been shown to result in photoreceptor cell degeneration (40). In order to determine whether aging had an effect on outer segment renewal, the phagosome content of the RPE was determined in rats of various ages at 1

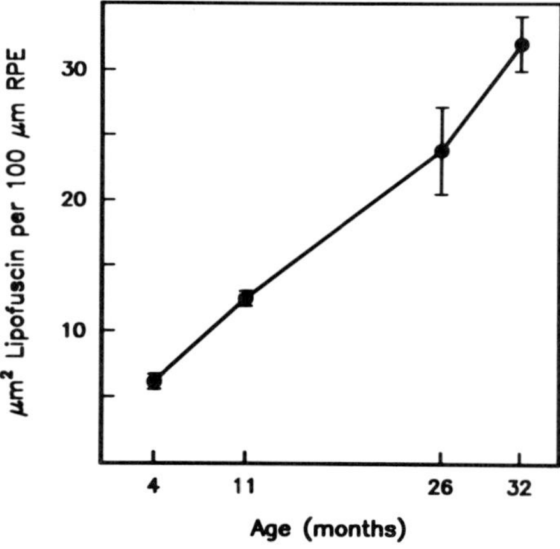

Fig. 2. Relationship between age and RPE lipofuscin content in pigmented ACI rats. The amount of lipofuscin in the RPE increases linearly during senescence. Adapted from Katz and Robison (27).

to 1.5 hours after light onset; the animals were maintained on a 12 hr/12 hr light/dark cycle throughout their lives. The phagosome content of the RPE declined dramatically with advancing age (27), suggesting that outer segment turnover slows during the aging process. Whether this change is due to a primary alteration in the RPE or in the photoreceptor cells themselves, or is secondary to a systemic effect of aging is currently unknown.

Besides participating in photoreceptor outer segment turnover, the RPE performs a number of other funcations that are essential in maintaining photoreceptor cell integrity. Among the most important of these functions are the transport and storage of vitamin A, one form of which is a constituent of the visual pigments. The fact that retinal photoreceptor cells degenerate in response to dietary viamin A deficiency (4,5) suggests that alterations in vitamin A metabolism by the RPE could be detrimental to the neural retina. In the rat, the RPE stores very little vitamin A when the animal is in a dark-adapted state. Light adaptation results in visual pigment bleaching and the release of vitamin A from the photoreceptor cells. The vitamin A is taken up by the RPE where it is stored as fatty acid esters until it is required for visual pigment regeneration. Recent investigations have suggested that the metabolism of vitamin A during light- and dark-adaptation is altered as a

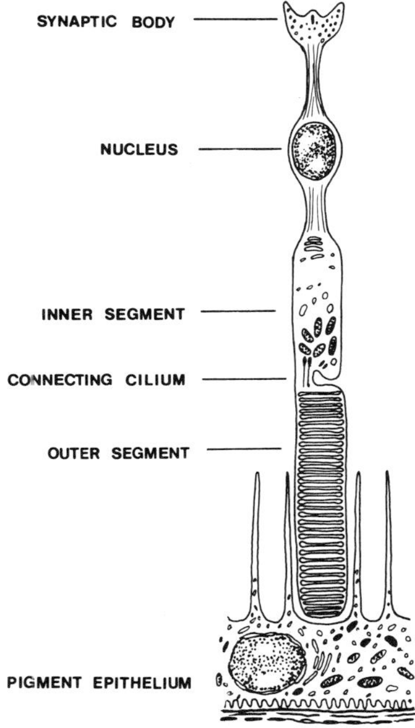

Fig. 3. Diagram illustrating the relationship between a rod photoreceptor cell and the retinal pigment epithelium. The rod outer segment consists of a stack of flattened membranous discs within the plasma membrane. New disc components are continually synthesized in the inner segment and assembled into discs in the region of the connecting cilium. This assembly process results in the displacement of the older discs toward the RPE. Periodically, packets of the oldest discs are shed from the outer segment tips and taken up by the RPE.

result of the aging process (30). The ratio of RPE to neural retina vitamin A content in light-adapted eyes increases as a function of age. In addition, the ratio of the vitamin A palmitate to stearate ester in the RPE becomes significantly elevated during senescence. The existence of these and other age-related alterations in vitamin A distribution and composition in the eye may reflect detrimental changes that could be responsible for some of the deterioration in visual function that occurs during aging.

In addition to participating in photoreceptor cell membrane renewal and in the delivery and storage of vitamin A for visual pigment synthesis, the RPE performs a number of other functions essential for maintaining photoreceptor cell integrity (55). Despite their importance in supporting visual capacity, most of these RPE functions have yet to be evaluated with respect to the effects of aging. Further characterization of the effects of senescence on the RPE should lead to a better understanding of the factors underlying the development of senile retinopathies.

NUTRITIONAL EFFECTS ON AGE-RELATED CHANGES OF THE RETINA AND RPE

In order to determine the mechanisms underlying age-related changes in the retina and its supporting tissues, experimental approaches will need to be adopted that will make it possible to identify the molecular and cellular events that produce the observed morphological and physiological alterations in these tissues. A number of theories regarding the primary mechanisms of mammalian aging have been proposed. Of these, the free radical theory (15,16) has been subjected to the greatest amount of experimental evaluation. According to this theory, many age-related alterations are secondary to the damaging effects of free radical-mediated reactions, primarily those involving oxygen-based radicals (autoxidation reactions). The significance of free radical damage in aging processes can be evaluated experimentally because it is possible, through dietary manipulation, to regulate the amount of free radical damage which occurs in vivo. By varying dietary intakes of antioxidants and pro-oxidants in experimental animals, it has been possible to study the role of oxygen-based radical damage in the development of age-related alterations in the retina and associated tissues.

Of the many molecular components of cells that are likely to be damaged by autoxidation, those reactions involving polyunsaturated lipids have been the most thoroughly studied. Perhaps the most important biological lipid antioxidant is vitamin E. Manipulation of dietary vitamin E content has been a useful tool for evaluation of the potential significance of lipid autoxidation in the age-related changes of the retina. Other dietary components important in regulating in vivo autoxidation rates include selenium, chromium, and the sulfur-containing amino acids. Dietary intake of these compounds has also been varied in order to determine whether autoxidation may be involved in age-related alterations in retinal structure and function.

Vitamin E deficiency has been shown to greatly accelerate the age-related accumulation of lipofuscin in the RPE (22,47,48). Rats maintained on a vitamin E-deficient diet for 26 weeks had almost 4 times as much RPE lipofuscin as did animals fed a vitamin E-adequate diet (22). The vitamin E deficiency-related pigment had fluorescence properties essentially identical to those of the age-related pigment (32), suggesting that both pigments are generated via similar mechanisms. Combined deficiencies in vitamin E, selenium, chromium, and sulfur-containing amino acids also resulted in a substantial acceleration of RPE lipofuscin deposition (25,33). Addition of chromium and methionine to this deficient diet partially prevented the increase in RPE lipofuscin deposition (25); this is further evidence that autoxidation is involved in RPE lipofuscin accumulation.

Possible substrates for lipofuscin formation in the RPE include polyunsaturated fatty acids, which are abundant in the photoreceptor outer segment disc membranes (9), and vitamin A compounds, which are highly unsaturated and easily oxidized. It has been possible, by regulating dietary intake of vitamin A, to evaluate the role of vitamin A autoxidation in RPE lipofuscin deposition. Robison and colleagues (48) reported that the acceleration of RPE lipofuscin deposition induced in rats by vitamin E deficiency could be partially blocked if the animals were also made deficient in vitamin A. This finding suggested that vitamin A autoxidation may be involved in RPE lipofuscin formation. Recent experiments have indicated, however, that the effect of vitamin A deficiency on RPE lipofuscin content occurs even in the presence of adequate dietary vitamin E intake (22). Thus, while it is clear that vitamin A can influence RPE lipofuscin deposition, it is uncertain whether this vitamin A effect occurs via an autoxidative mechanism. Recently it has been observed that a lipofuscin-like autofluorescence develops in the degenerating photoreceptor cells of RCS rats with hereditary retinal dystrophy (21). The intensity of the autofluorescence was reduced substantially if the animals were fed a vitamin A-deficient diet (24) (Fig. 4). This finding provides further support for a role of vitamin A in the formation of RPE lipofuscin fluorophores. Additional evidence that vitamin A can influence RPE lipofuscin accumulation has been reviewed recently by Robison and Katz (46).

The effects of dietary manipulation on various features of the aging process in the retina, in addition to RPE lipofuscin accumulation, have been examined in a number of experiments. These investigations have focused almost exclusively on determining whether autoxidation is involved in the development of age-related alterations in the retina and associated tissues. A substantial number of similarities have been observed between the effects of aging and antioxidant nutrient deficiencies on the retina and RPE (31). For example, the age-related decrease in photoreceptor cell density that occurs in rats is greatly accelerated by deficiencies in antioxidant nutrients (22,25,48). While this observation is consistent with the possibility that autoxidation may be involved in the slow loss of photoreceptors during senescence, it does not exclude the potential that other mechanisms may be involved. Numerous pathological conditions, unrelated to the aging process, can also lead to photoreceptor cell death (1,5,40,42)

The existence of additional similarities between the effects of aging and antioxidant nutrient deficiency on the retina and RPE increase the likelihood that autoxidation may be involved in the senescent changes that occur in these tissues. Both aging and antioxidant nutrient deficiency result in an apparent reduction in photoreceptor outer segment phagocytosis by the RPE (25,27). This suggests that aging and antioxidant deficiency have similar effects on the outer segment renewal process. A pleomorphism in RPE cell size and shape, similar to that which develops in humans during aging, can be induced in rats by antioxidant nutrient deficiencies (25). Such deficiencies also result in an acceleration of the age-related deposition of lipid bodies in the RPE (25).

In addition to inducing changes in retinal morphology similar to those which occur during senescence, antioxidant nutrient deficiency also appears to accelerate age-related alterations in vitamin A metabolism in the retina-RPE complex. If rats are exposed to lights that produce almost complete visual pigment bleaches, most of the vitamin A in these tissues is deposited in the RPE as reintyl esters. In the rat, these esters consist almost exclusively of palmitate and stearate. As indicated previously, there is a progressive age-related increase in the ratio of retinyl palmitate to retinyl stearate in the rat RPE (27,30). This increase can be accelerated by feeding rats vitamin E-deficient diets (23), suggesting that both autoxidation and aging have similar effects on vitamin A metabolism. Not only is the composition of vitamin A esters in the RPE of light-adapted rats altered during aging,

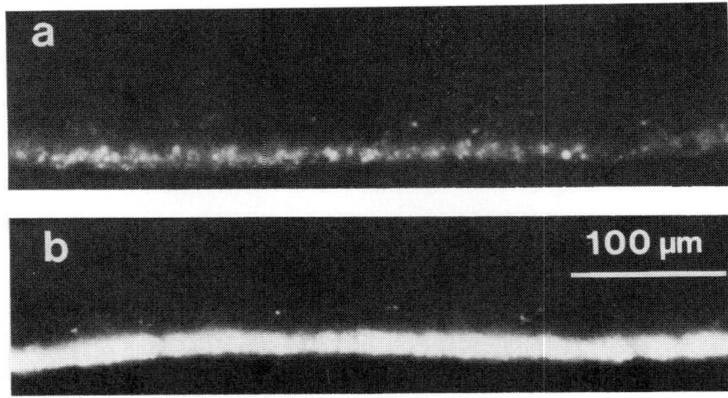

Fig. 4. Fluorescence micrographs of retinas from 66-day-old RCS dystrophic rats which had been fed either a retinol-deficient (a) or a retinol-containing diet (b) since weaning. The fluorescence intensity of the degenerating photoreceptor outer segment material is highly dependent on dietary retinol content.

but there is also an age-related increase in total RPE vitamin A content (30), despite the fact that retinal visual pigment content decreases during senescence (29). Again, antioxidant nutrient deficiency has similar effects; vitamin E deficiency results in an elevation of total RPE retinyl ester content in light-adapted animals (23), while reducing retinal visual pigment content (26). The similarities between the effects of aging and antioxidant nutrient deficiencies on vitamin A metabolism in the eye are consistent with the possibility that autoxidation may be involved in the alteration of vitamin A metabolism during senescence.

SIGNIFICANCE OF NUTRITIONAL EXPERIMENTS FOR HUMAN VISUAL FUNCTION

In order to develop an effective means of preventing human senile retinopathies, it will be necessary to gain a better understanding of the cellular and molecular events that underlie these diseases. The strong correlation between chronological age and the incidence of retinal pathologies indicate that such pathologies probably develop as a consequence of the normal aging process. Therefore, it is essential, if we are to control these diseases, that we gain a better understanding of how senescence affects retinal structure and function.

A beginning has been made in characterizing age-related changes in the rtina and associated tissues, but the association between these changes and the development of retinal disease is at present still largely a matter of speculation. For example, retinal neuron dropout appears to be a concomitant of the aging process, but whether this cell loss represents an actual decrement in retinal function or merely the elimination of redundancy is unknown. Likewise, the effects of lipofuscin accumulation on RPE function remains to be determined. Until we can establish the nature of the relationship between simple age-related alterations in the retina and frank pathological changes, we will be unable to design a rational approach for the prevention of senile retinopathies.

The discovery through dietary experiments that autoxidation can accelerate many age-related alterations in the retina and associated tissues not only has led to a better understanding of the mechanisms underlying senescent changes in these tissues, but also may allow us to characterize the association between these aging changes and senile retinopathies. If an appropriate animal model for AMD could be found, it would be possible to determine whether dietary manipulations which accelerate normal aging changes also produce an increase in the incidence of AMD at earlier ages. Such experiments would be of great assistance in designing a program for the prevention of AMD.

SUMMARY AND CONCLUSIONS

Age-related diseases of the retina are serious health problems in developed countries, affecting 8 to 10% of the population between 65 and 74 years of age (11,34), and almost 30% of those 75 to 85 years old (34) in the United States. Unless effective means of preventing senile retinopathies are discovered, the social significance of these diseases is likely to increase in the future, as the proportion of the elderly in the populations of developed countries continues to increase. Research is currently under way that should lead to a better understanding of the molecular and cellular bases for these diseases. It is hoped that such understanding will ultimately lead to an effective means of prevention.

Nutritional experiments performed on experimental animals have established a link between potentially damaging autoxidation reactions and the development of normal age-related changes in the retina and associated tissues. It remains to be determined whether these reactions are involved in age-dependent changes that are recognized as being pathological. If such an association can be established, it would be reasonable to investigate the possibility of enhancing the normal antioxidant protective mechanisms of the retina, and thereby possibly retarding the development of senile retinopathies.

ACKNOWLEDGEMENTS

Work conducted in the laboratory of M.L. K. and reviewed in this chapter was supported in part by a research grant EY 06458 from the National Eye Institute, and by an unrestricted grant from Research to Prevent Blindness, Inc. Our thanks to Garry Handelman for his helpful comments on the manuscript, and to Sherry DiMaggio for assistance in manuscript preparation.

REFERENCES

1. G. Aguirre and L. Rubin, Diseases of the retinal pigment epithelium in animals, 1979, in: "The Retinal Pigment Epithelium," K.M. Zinn and M.F. Marmor, eds., Harvard University Press, Cambridge, MA.

2. A.G. Balazsi, J. Rootman, S.M. Drance, M. Schulzer, and G.R. Douglas, 1984, The effect of age on the nerve fiber population of the human optic nerve, Am. J. Ophthalmol. 97:760.

3. K.T. Brown, 1969, The elctroretinogram: its components and their origins, in: "The Retina: Morphology, Function, and

Clinical Characteristics," B.R. Straatsma, M.O. Hall, R.A. Allen, and F. Crescitelli, eds., University of California Press, Berkeley.

4. L. Carter-Dawson, T. Kuwabara, P.J. O'Brien, and J.G. Bieri, 1979, Structural and biochemical changes in vitamin A-deficient rat retinas, Invest. Ophthalmol. Vis. Sci. 18:437.

5. J.E. Dowling, 1964, Nutritional and inherited blindness in the rat, Exp. Eye Res. 3:348.

6. L. Feeney, 1978, Lipofuscin and melanin of the human retinal pigment epithelium, Invest. Ophthalmol. Vis. Sci. 17:583.

7. L. Feeney, J.A. Grieshaber, and M.J. Hogan, 1965, Studies on human macular pigment, in: "The Structure of the Eye II. Symposium," J.W. Rohen, ed., F.K. Schattauer-Verlag, Stuttgart.

8. L. Feeney-Burns, E.S. Hilderbrand, and J. Eldridge, 1984, Aging human RPE: morphometric analysis of macular, equitorial, and peripheral cells, Invest. Ophthalmol. Vis. Sci. 25:195.

9. S.J. Fliesler and R.E. Anderson, 1983, Chemistry and metabolism of lipids in the vertebrate retina, Prog. Lipid Res. 22:79.

10. E. Friedman and M. Ts'o, 1968, The retinal pigment epithelium II. Histologic changes associated with age. Arch. Ophthalmol. 79:315.

11. J.P. Ganley and J. Roberts, 1983, "Eye Conditions and Related Need for Medical Care Among Persons 1-74 Years of Age: United States 1971-72," DHHS Pub. No. (PHS) 83-1678, U.S. Government Printing Office, Washington.

12. S. Gartner and P. Henkind, 1981, Aging and degeneration of the human macula. 1. Outer nuclear layer and photoreceptors, Br. J. Ophthalmol. 65:23.

13. H.J. Glatt and P. Henkind, 1979, Aging changes in the capillary bed of the rat retina, Microvascular Res. 18:1.

14. A. Haas, J. Flammer, and U. Schneider, 1986, Influence of age on the visual fields of normal subjects, Am. J. Ophthalmol. 101:199.

15. D. Harman, 1956, Aging - a theory based on free radical and radiation chemistry, J. Gerontol. 11:298.

16. L. Hayflick, 1985, Theories of biological aging, Exp. Gerontol., 20:145.

17. M.J. Hogan, 1967, Bruch's membrane and disease of the macula: role of elastic tissue and collagen, Trans. Ophthalmol. Soc. U.K. 87:113.

18. M. J. Hogan, 1972, Role of the retinal pigment epithelium in macular disease, Trans. Amer. Acad. Ophthalmol. Otolaryngol. 76:64.

19. M.J. Hogan and J. Alvarado, 1967, Studies on the human mucula. IV. Aging changes in Bruch's membrane, Arch. Ophthalmol. 77:410.

20. J.L. Jay, R.B. Mammo, and D. Allan, 1987, Effect of age on visual acuity after cataract extraction, Br. J Ophthalmol. 71:112.

21. M.L. Katz, C.M. Drea, G.E. Eldred, H.H. Hess, and W.G. Robison, Jr., 1986, Influence of early photoreceptor degeneration on lipofuscin in the retinal pigment epithelium, Exp. Eye Res. 43:561.

22. M.L. Katz, C.M. Drea, and W.G. Robison, Jr., 1986, Relationship between dietary retinol and lipofuscin in the retinal pigment epithelium, Mech. Age. Dev. 35:291.

23. M.L. Katz, C.M. Drea, and W.G. Robison, Jr., 1986, Dietary vitamins A and E influence retinyl ester content and composition in the retinal pigment epithelium, Biochim. Biophys. Acta. 924:432.

24. M.L. Katz, G.E. Eldred, and W.G. Robison, Jr., 1987, Lipofuscin autofluorescence: evodence for vitamin A involvement in the retina, Mech. Age. Dev. 39:81.

25. M.L. Katz, K.R. Parker, G.J. Handelman, T.L. Bramel, and E.A. Dratz, 1982, Effects of antioxidant nutrient deficiency on the retina and retinal pigment epithelium of albino rats: a light and electron microscopic study, Exp. Eye Res., 34:339.

26. M.L. Katz, K.R. Parker, G.J. Handelman, C.C. Farnsworth, and E.A. Dratz, 1982, Structural and biochemical effects of antioxidant nutrient deficiency on the rat retina and retinal pigment epithelium, Ann. N.Y. Acad. Sci. 393:196.

27. M.L. Katz and W.G. Robison, Jr., 1984, Age-related changes in the retinal pigment epithelium of pigmented rats, Exp. Eye Res. 38:137.

28. M.L. Katz and W.G. Robison, Jr., 1985, Senescence and the retinal pigment epithelium: alterations in basal plasma membrane morphology, Mech. Age. Dev. 30:99.

29. M.L. Katz and W.G. Robison, Jr., 1986, Evidence of cell loss from the rat retina during senescence, Exp. Eye Res. 42:293.

30. M.L. Katz and W.G. Robison, Jr., 1987, Age-related alterations in vitamin A metabolism in the rat retina and retinal pigment epithelium, Exp. Eye Res. 44:939.

31. M.L. Katz, W.G. Robison, Jr., and E.A. Dratz, 1984, Potential role of autoxidation in age changes of the retina and retinal pigment eipithelium of the eye, in: "Free Radicals in Molecular Biology, Aging, and Disease," D. Armstrong, R.S. Sohal, R.G. Cutler, and T.S. Slater, eds., Raven Press, New York.

32. M.L. Katz, W.G. Robison, Jr., R.K. Herrmann, A.B. Groome, and J.G. Bieri, 1984, Lipofuscin accumulation resulting from senescence and vitamin E deficiency: spectral properties and tissue distribution, Mech. Age. Dev. 25:149.

33. M.L. Katz, W.L. Stone, and E.A. Dratz, 1978, Fluorescent pigment accumulation in retinal pigment epithelium of anitoxidant-deficient rats, Invest Ophthalmol. Vis. Sci. 17:1049.

34. M.M. Kini, H.M. Liebowitz, T. Colton, R.J. Nickerson, J. Ganley, and T.R. Dawber, 1978, Prevalence of senile cataract, diabetic retinopathy, senile macular degeneration, and open-angle glaucoma in the Framingham eye study, Am. J. Ophthalmol. 85:28.

35. A.L. Kornzweig, 1979, Aging of the retinal pigment epithelium, in: "The Retinal Pigment Epithelium," K.M. Zinn and M.F. Marmor, eds., Harvard University Press, Cambridge, MA.

36. T. Kuwabara, 1979, Age-related changes of the eye, in: "Special Senses and Aging, A Current Assessment," S.S. Han and D.H. Coons, eds., University of Michigan Press, Ann Arbor.

37. T. Kuwabara, J.M. Carroll, and D.G. Cogan, 1961, Retinal vascular patterns. Part III. Age, hypertension, absolute glaucoma, injury, Arch. Ophthalmol. 65:708.

38. Y. Lai, R.O. Jacoby, and A.M. Jones, 1978, Age-related and light-associated retinal changes in Fisher rats, Invest. Ophthalmol. Vis. Sci. 17:634.

39. M.M. LaVail, 1976, Rod outer segment disc shedding in relation to cyclic lighting, Exp. Eye Res. 23:277.

40. M.M. LaVail, 1979, The retinal pigment epithelium in mice and rats with inherited retinal degeneration, in: "The Retinal Pigment Epithelium," K.M. Zinna and M.F. Marmor, eds., Harvard University Press, Cambridge, MA.

41. M.M. LaVail, 1980, Circadian nature of rod outer segment disc shedding in the rat, Invest. Ophthalmol. Vis. Sci. 19:407.

42. M.M. LaVail, J.G. Hollyfield, and R.E. Anderson, eds., 1985, "Retinal Degeneration: Experimental and Clinical Studies," Alan R. Liss, New York.

43. H. Mishima, H. Hasebe, and K. Kondo, 1978, Age changes in the fine structure of the human retinal pigment epithelium, Jpn. J. Ophthalmol. 22:476.

44. M. Nagata, M.L. Katz, and W.G. Robison, Jr., 1986, Age-related thickening of retinal capillary basement membranes, Invest. Ophthalmol. Vis. Sci., 27:437.

45. J.M. Ordy, K.R. Brizzee, and J. Hansche, 1980, Visual acuity and foveal cone density in the retina of the aged rhesus monkey, Neurobiol. Aging 1:133.

46. W.G. Robison, Jr. and M.L. Katz, 1987, Vitamin A and lipofuscin, in: Cell and Developmental Biology of the Eye, the Microenvironment and Vision," T. Sheffield and S.R. Hilfer, eds., Springer-Verlag, New York.

47. W.G. Robison, Jr., T. Kuwabara, and J.G. Bieri, 1979, Vitamin E deficiency and the retina: photoreceptor and pigment epithelial changes, Invest. Ophthalmol. Vis. Sci. 18:683.

48. W.G. Robison, Jr., T. Kuwabara, and J.G. Bieri, 1980, Deficiencies of vitamins E and A in the rat. Retinal damage and lipofuscin accumulation, Invest. Ophthlamol. Vis. Sci. 19:1030.

49. S.H. Sarks, 1976, Aging and degeneration in the macular region: a clinico-athological study, Br. J. Ophthalmol. 60:324.

50. N.L. Shinowara, E.D. London, and S.I. Rapoport, 1982, Changes in retinal morphology and glucose utilization in aging albino rats, Exp. Eye Res. 34:517.

51. B.W. Streeten, 1961, The sudanophilic granules of the human retinal pigment epithelium, Arch. Ophthalmol. 66:125.

52. I. Weisse, H. Stoetzer, and R. Seitz, 1974, Age-and light-dependent changes in the rat eye, Virchows Arch. A Pathol. Anat. Histol. 362:145.

53. J.J. Weiter, F.C. Delori, G.L. Wing, and K.A. Fitch, 1986, Retinal pigment epithelial lipofuscin and melanin and choroidal melanin in human eyes, Invest. Ophthalmol. Vis. Sci. 27:145.

54. G.L. Wing, G.C. Blanchard, and J.J. Weiter, 1978, The topography and age relationship of lipofuscin concentration in the retinal pigment epithelium, Invest. Ophthalmol. Vis. Sci. 17:601.

55. K.M. Zinn and M.F. Marmor, eds., 1979, "The Retinal Pigment Epithelium," Harvard University Press, Cambridge, MA.

THE INFLUENCE OF ANTICHOLINESTERASE PESTICIDES ON VISUAL MOTOR FUNCTION: A POSSIBLE ACCELERATOR IN AGING

Satoshi Ishikawa

Department of Ophthalmology, School of Medicine
Kitasato University
Sagamihara, Kanagawa, Japan

ABSTRACT

Farmers exposed to organophosphorus pesticides (OP) for over 10 years and children with environmental exposure to OP for over 3 years manifested balance disturbance and abnormal constriction of the pupil.

When body balance and pupil involvement were quantitatively compared with age-matched controls, both impairments were significantly advanced and suggest a possible acceleration in the aging process.

INTRODUCTION

Organophosphorus pesticides (OP) have been extensively used throughout the world and, in many cases, acute intoxication has been reported [1]. OP exposure produces an intense cholinomimetic action in cholinergically innervated nervous systems. Therefore, diagnosis of acute intoxication is very easy. However, with chronic OP exposure, diagnosis is not as simple. Chronic exposure occasionally produces "delayed neurotoxicity" in humans [2,5]. This is summarized as "optic-autonomic-neuropathy" [6]. On the other hand, patients often complain of severe dizziness, oscillopsia, difficulty looking at near distances, a dimness or darker appearance of the visual field, and difficulty in standing. The dizziness improves when the patient follows a smoothly moving target and the reverse is true for eye closure. This suggests involvement of the vestibular apparatus related to balance.

In the present study, we will first introduce a typical patient with chronic OP intoxication. Special emphasis on his standing ability and his ability related to eye movement loading, i.e., smooth pursuit eye movement, will be examined. Secondly, the standing ability of subjects who were diagnosed with chronic organophosphorus intoxication and the results obtained from the control group will be shown. Thirdly, the pupil findings of 20 hospitalized patients with chronic organophosphorus intoxication will be discussed. Finally, the methods of diagnosis will be emphasized, considering the aging of the body balance and pupil systems.

The Effects of Aging and Environment on Vision
Edited by D.A. Armstrong *et al.*, Plenum Press, New York, 1991

METHOD

Balance Study

109 patients who had visited our department during the past five years with chronic OP intoxication were studied. The mean age was 35.8 years, and the age range was from 6 to 58 years. Over 55% of the patients were farmers with at least 10 years history of spraying with pesticides. The rest of the patients were adults and children whose parents were farmers. The history of environmental exposure to OP in children was over 3 years. They were diagnosed using the following criteria [7]:

1. Prolonged or heavy contact with OP
2. Ocular involvements
   a. Nerve fiber bundle defect at maculo-papillary fibers
   b. Dark adaptation
   c. Smooth pursuit movement
   d. Pupil and accommodation
3. Peripheral sensorial neuropathy
4. Involvement with other autonomic nervous system functions resulting in the following:
   a. Gastro-intestinal symptoms
   b. Nausea
   c. Fatigability
   d. Numbness of the leg or hand
   e. General malaise
   f. Muscle pain
   g. Headache
   h. Visual discomfort, such as reduced vision and/or difficulty of focusing either at far or near distances.
   i. Dizziness
   j. Involuntary twitching or burning sensation of the leg or arm.

The patients had to answer 70 questions on a written questionnaire. when there were over 20 positive answers, special examinations were conducted [5].

After the administration of antidotes, the following were noted: a reduction in erythrocyte cholinesterase activity, an improvement in the clinical manifestations, and an elevation in acetylcholinesterase.

Four major pesticides included in this study were:

1. Fenitrothion (Sumithion:0, 0-dimethyl-0/3-methyl-4 nitrophenyl/ phosphrothionate)
2. 0,0 diethyl 0-(2-isopropyl 6-methyl-4-pyrimidinyl) phosphorothioate, diazinon.
3. Dipteryx, Trichlorfon: 0,0-dimethyl (2,2,2-trichloro-1-hydroxyethl) phosphorate
4. 0,0 dimethyl -1-2,2-dichlorovinyl phosphate(dichlorvos).

The control group in the standing ability study consisted of 1000 healthy subjects of varying ages with no contact with OP. In the balance study of the loading of smooth pursuit eye movement, 32 healthy controls, with mean age of 32.0 years and no contact with OP, were randomly selected.

Standing ability was examined using the following procedures. The subject stood on the plate of a electrogravitiography (EGG-Anima) and fixed his eyes on a projected target 1.5 meters away. After this, smooth pursuit eye movement, and saccadic eye movement, were elicited. The target was sinusoidally moved in the horizontal plane at frequencies ranging from 0.3 to 0.7 Hz, with an amplitude from 0 degrees (no eye movement with central fixation) to 40 degrees. The subject was asked to follow the target. The center of gravity at the sole was recorded by EGG in both right/left (X) and front/back (Y). Each examination took 20 seconds. The effect of loading on smooth pursuit tracking on EGG was also studied.

Pupil study

1. Pupil area against age: Pupils were examined in darkened room immediately after dark adaptation of 15 minutes by infrared video-pupillography(Hamamatsu photonics C-2515)[8]. Twenty OP patients hospitalized for treatment were selected out of the 109 patients. The clinical manifestation of age, cholinesterase activity, OP, and pupil of the patients are shown in Table I. The pupil area in the dark was compared with those from the 100 healthy controls who were selected from University Hospital employees.

2. Photo stress study of the pupil: The same pre-dark adaptation was given on a separate day with 1. examination. After measurement of the pupil area, a photo stress with light intensity of 15,000 cd/m for 20 seconds was given to the fovea by indirect binocular ophthalmoscopy, with 20 diopters lens viewing the macula. Immediately after the photo stress, recovery of the pupil area was examined and compared with the normal controls, who had no contamination with OP. The pupil was measured one minute and three minutes afterward. Pupil areas against time in minutes were measured and compared with 20 approximately age matched controls. These were selected from 32 healthy subjects, with mean age of 35.7 years.

TABLE I
Age, cholinesterase activites and blood level of organophosphorus pesticides (OP)

NO.	Age	Cholinesterase Activity Red Cell (umol/ul/min)	Serum	OP (ppb)	Detected OP(name)
1	43	1.8	5.0	5.4	D (Diazinon)
2	61	1.8	7.5	34.2	S (Sumithion)
3	38	1.8	7.4	27.1	S
4	51	2.1	6.6	6.7	DDVP(Dichlorphos)
5	21	1.8	4.6	27.7	D
6	56	1.6	8.0	38.1	D
7	72	1.8	7.4	8.3	D
8	47	2.0	7.2	14.2	S
9	47	1.4	7.3	35.1	D
10	75	1.9	5.0	24.9	D
11	52	2.0	5.6	51.8	S
12	53	1.4	6.3	24.9	DDVP
13	33	1.2	5.7	253.1	S
14	49	1.6	5.0	28.1	D
15	71	1.7	5.6	79.6	S
16	60	1.6	5.3	53.1	S
17	53	1.8	6.6	19.1	S
18	78	1.9	5.4	18.7	DDVP
19	54	1.6	4.7	29.9	S
20	40	1.2	3.0	48.7	S
Mean	52.7	1.7	6.0	41.4	
S.D.	14.0	0.25	1.18	52.9	

LABORATORY EXAMINATIONS

Detailed descriptions have been described elsewhere [9]. Initially, erythrocyte cholinesterase (true-ChE) and serum ChE(pseud-ChE) were measured (Ellman) [10]. Normal value in our laboratory is 1.8 - 2.2 umol/ml/minute in red cell or true ChE and 4.5 - 6.5 umol/ml/minute in serum or pseud ChE, respectively, Measurements from most of the patients revealed that the former was reduced and the latter rather elevated [11]. This is an important contrast between chronic and acute OP intoxication, since both ChE activities are reduced in acutely intoxicated patients. Routine blood chemistry examinations of the liver, such as GOT or GPT, were meaningless in establishing the diagnosis of chronic OP intoxication. However, there are many publications available describing OP as safe pesticides.

Studying the following items in the blood may assist in the diagnosis of chronic OP patients. These are a reduction of vitamin C, folic acid, selenium, zinc, and magnesium levels. an abnormal level of glutathione peroxidase and superoxide dismutase may also help in the diagnosis.

The residue of OP in the urine or in the blood occasionally helps in the diagnosis; however, reductions of the residue of OP after the use of the antidote is even more reliable in establishing the diagnosis. This is also applicable to erythrocyte ChE elevation. For brevity in this text, the results of the blood chemistry study will not be mentioned, except for that of ChE.

For example, one 48 year old male the highest EGG sprayed DDVP about 11 times per month. His presumed daily intake was 2.05mg/Kg, calculated from sprayed DDVP. This is about 70 times the maximum allowable daily intake of 0.03mg/Kg. Clinically, the patient had severe peripheral neuropathy. He had difficulty looking at near distances for several years. No hypermetropia existed. In order to obtain objective findings from these intoxicated patients, the balance and pupil study were performed.

RESULTS AND DISCUSSION

Balance study with smooth pursuit loading

Typical results obtained from the patients are given in Figure 1. The inhibitory effect o smooth pursuit on EGG is shown. The patient had been a professional OP sprayer for the past 18 years and had a severe "optic-autonomicperipheral neuropathy". The effects of smooth pursuit stimulus at 0.5 Hz and different amplitudes (0-40 degrees) on EGG(X & Y) and no EGG-trajectory are given. He could follow the target at 0 degrees; otherwise, no eye movement was elicited. His standing ability was highly disturbed. This could be clearly seen in the EGG and the trajectory of the EGG. When the horizontal eye movement stimulus gradually increased till 40 degrees, his balance improved. This inhibition of body sway by loading os smooth pursuit eye movement was a unique phenomenon seen in most of the patients. The inhibition was not seen when the patient was asked to look at an optokinetic drum rotation at varying velocities while in his primary position When the horizontal extraocular muscles on one eye were paralyzed by retrobulbar injection (1.5ml of 2% procaine into muscle belly of the right lateral and medial rectus muscles), the above inhibition was not seen. Therefore, this inhibition was not only due to sensorial for example, corollary discharge, but was due to proprioceptive effect possibly from the extraocular muscles to the vestibular system, affected by OP where controls the body balance. The muscle spindle, as well as the palisade ending at the muscletendinous junction of the extraocular muscle is innervated by the

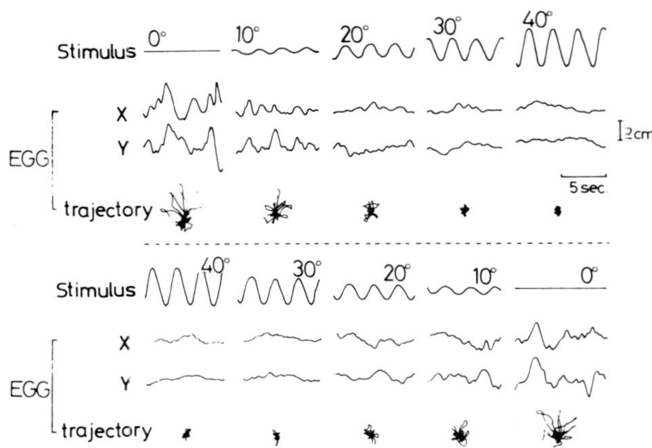

Figure 1. Smooth pursuit of the eyes: A marked improvement of standing ability with the loading of smooth horizontal eye movement. Stimulus: target movement (0.5 Hz) from 0 degrees (no eye movement loading) till 40 degrees amplitude. EGG: electro gravitiography.

cholinergic nerve. This inhibition was also seen in the patients with myasthenia gravis who had been treated by anticholinesterase drugs for at least 6 months with carbamate ester. The inhibition was never seen from other neurological diseases. Therefore, this sign was considered to be a specific phenomenon in patients with chronic OP intoxication.

The above inhibition was also observed during saccadic movement in both normal and OP patients. This test is especially useful for screening the patients. The loading of ocular smooth pursuit movement on EGG was examined in all patients and controls. The results are given in Figure 2 as an effect of loading by ocular smooth pursuit movement on EGG in both patients and normal controls. The results were classified into 4 types: a. Inhibited EGG (66 cases) b. Excited (29 cases) c. Combined findings (5 cases) d. No Changes (6 cases) in the patients. Therefore, 61.1% of chronic OP intoxication patients could be diagnosed by this method. Type a. and type b. improved with oral administration of antidotes, such as pralidoxime methiodide 1 gram/day and/or atropine methyl-nitrate 0.5mg/day, for a prolonged period of time. It was evident that the improvement of standing ability by loading with the smooth pursuits ocular stimulus was due to anticholinesterase action evoked by OP, mostly at the brain stem where abundant acetylcholinesterase exists.

Standing ability and age

Standing ability (i.e., the area of EGG cm2 obtained from 20 patients with OP and those from the controls with eyes open), against age in years is shown in Figure 3. Closed circles with downward vertical bars denote the

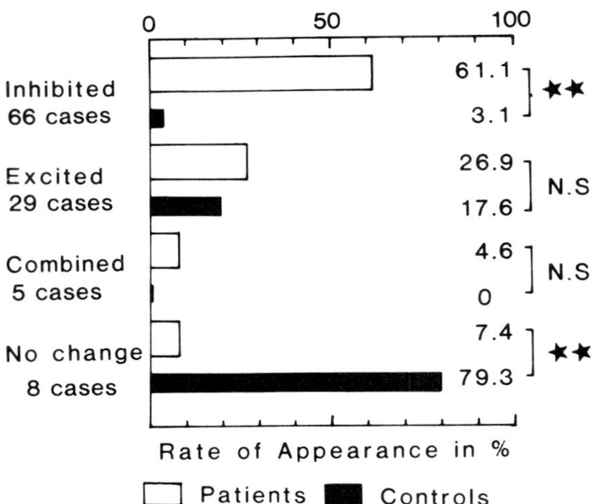

Figure 2. Effect of loading by ocular smooth pursuit movement on EGG. Comparison between patients (white squares) and controls (solid squares). There are 4 types of responses on EGG by loading of the eye movement: inhibited, excited, combined and no change. Significant differences existed for inhibited and no change, $p<0.01$.

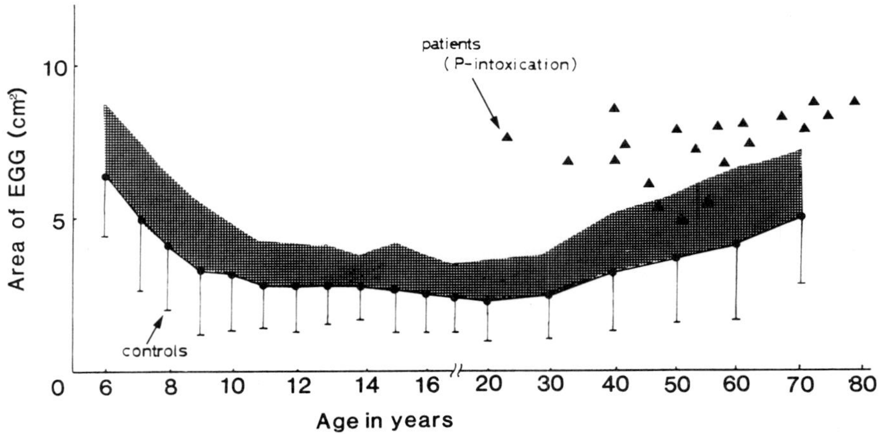

Figure 3. Area of EGG (both eyes open) against age in years in both controls (solid circles) and OP patients (solid triangles). The means and the standard deviations of the controls are shown. The area of EGG is higher in most of hospitalized patients (triangles) with chronic organophosphorus intoxication.

means and standard deviations of the normal controls in each age group. The same is true with upward standard deviations (shade area). The EGG obtained from 20 patients with OP is plotted against age. The EGG of patients are larger than that of the controls, except in three cases (No. 8, 11 & 17) where true cholinesterase activities were 2.0, 2.0 and 1.8. This was slightly higher than the other patients. Increased EGG is more obvious in younger ages. This means that the standing ability of the patients is more involved and even at age of 20, the patients mean value was 6.68 +/- 1.98 cm2. Even at a young age, the area is closer to that of a 60-70 year old, indicating involvement of the vestibulo-ocular control system in the brain stem. The possibility of acceleration in the aging process is suggested.

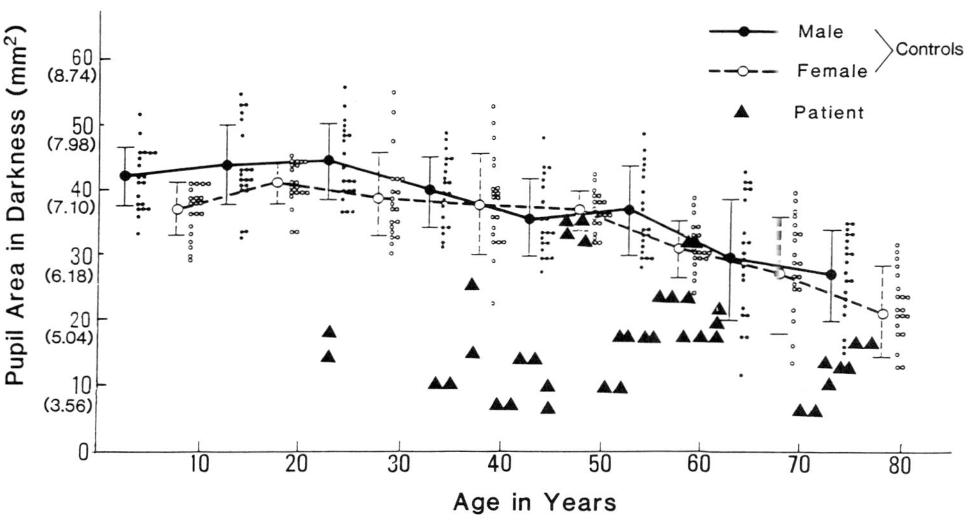

Figure 4. Pupil area in darkness is mm2 against age in years of both controls (male: solid circles, female: open circles) and patients (both eyes:triangles). The pupils of the patients were generally smaller than those of the controls in all age groups except in 3 cases.

Pupil area against age

The patients with a larger area of EGG over the mean value were selected and underwent further pupil study. The pupil area in darkness showed scattered results among normal individuals at all ages (males:solid circles, females: open circles). Never the less, an age trend could be discerned and it became clearer when the measurements of relatively large groups of subjects were averaged. Pupil area in darkness against age in years is plotted together with the controls and the OP patients, and the results are shown in Fig. 4. Twenty patients with OP (triangles), demonstrated smaller pupils than those of the controls in all age groups except for three cases(47, 47 and 58 years).

Thus, the patient's pupil is constricted in darkness. However, constricted pupil is not a specific sign of the OP patient, especially in chronic intoxication. Very often, pupil size is normal or rather dilated in a child in moderately illuminated room [12]. Therefore, the following test should be performed to establish the diagnosis.

<u>Photo stress test of the pupil</u>

The pupil sized elicited by intense light stimulus in the OP patients are unique. The summary of the photo stress test in the OP patients and the controls are shown in Table II. All measurements were done in a dark room by infrared pupillography. In the control group, the pupil returned to regular size even 3 minutes after the photo stress. When averaged, the results from the controls were: before photo stress: 32.34, 1 minute after photo stress 15.38, and 3 minutes after photo stress 30.36 mm^2, respectively. However, the pupils of the patients were: 17.80, 4.92, and 9.15 mm^2, respectively.

Table II
Pupil area (mm^2) in darkness by photo stress test
(Mean and S.D.)

	Patients	Controls	Difference
Before photo stress	17.80 + 8.14	32.34 + 10.13	$P<0.001$
1 minute after photo stress	4.92 + 3.14	15.38 + 7.69	$P<0.001$
3 minutes after photo stress	9.15 + 5.53	30.36 + 8.31	$P<0.001$

Pupil area before, 1 minute after, and 3 minutes after the photo stress test in patients and controls. Significant constriction of the pupils can be seen in patients, especially after the photo stress.

The pupils of the patients were smaller in the darkness even before photo stress but with the variances. Pupillary constriction evoked by photo stress (ps) produced an intense constriction of the pupils, especially in the patients seen one minute after photo stress ($p<0.001$). This constricted condition lasted up to 3 minutes in the patients ($p<0.001$). If a patient's pupil is still over 50% constricted at 3 minutes after the photo stress, the clinical sign is generally very severe.

We routinely use this test as an aid in establishing diagnosis. This intense miosis can be explained by over accumulation of acetylcholine at the sphincter receptor followed by inhibition of ChE by OP.

In conclusion, chronic environmental exposure to OP is very dangerous because it produces various effects on the eye and brain stem, where abundant cholinesterase exists. As shown in this study, balance and pupil examination are highly recommended to establish to diagnosis. Chemical measurement of the blood should follow. We would like to emphasize the early discovery and immediate treatment of the patient is absolutely necessary for farmers who use OP pesticides for prolonged periods of time. If left untreated and undiagnosed, some patients may lose their vision and experience severe neuropathy in their lower legs. Finally environmental exposure to OP pesticides may be an accelerator in aging.

REFERENCES

1. Grob, D. Harvey, A.M., 1953, The effect of nerve gas poisoning. Amer. J. Med. 14:52-58.

2. Ishikawa, S., 1971, Eye injury by organic phosphorus insecticides-preliminary report. Japn. J. Ophthalm. 15:60-68.

3. Ishikawa, S., 1973, Chronic optico-neuropathy due to environmental exposure of organophosphate pesticides (Saku Disease)-Clinical and experimental study. Acta. Soc. Ophthalm. Jap. 77:1835-1886.

4. Ishikawa, S., Miyata, M., Development of mypoia following chronic organophosphate pesticide intoxication: An epidemiological and experimental study. "Neurotoxicity of the Visual System" (Eds. W. H. Merignan & B. Weiss) pp 233-254, Raven Press, New York.

5. Suzuki, H., Ishikawa, S., 1974, Ultrastructure of the ciliary muscle treated by organophosphate pesticide in Beagle Dogs. J. Ophthal. 58:931-940.

6. Ishikawa, S. and Ohto, K., 1972, Opticneuropathy induced by environmental exposure to organophosphate pesticides. Proceeding of 5th Afro-Asian Congress of Ophthalmology, (Tokyo) pp 434-444.

7. Ishikawa, S., Ozawa, H., and Miyata, M., 1983, Abnormal standing ability in patients with organophosphate pesticide intoxication (chronic cases). Agressologie 24:74-75.

8. Ishikawa, S., Naito, M., and Inaba, K., 1970, A new videopupillography. Ophthalmologica: 160:248-259.

9. Ishikawa, S. and Ozawa, H., 1983, Disturbed balance in chronic organophosphate intoxication. Proceedings of 7th Int. Symp. Int. Soc. Posturography, Houston, Texas. pp 295-301.

10. Ellman, G.L., Courtney, K.D., 1961, A new and rapid colormetric determination of acetylcholinesterease. Biochem. Pharmacol., 7:88-92.

11. Davey, P.G. and Shearer, R.W., 1986, Hypersensitivity caused by environmental chemicals in particular pesticides. Clinical Ecology 4:35-39.

12. Hasegawa, S. and Ishikawa, S., 1989, Age changes of pupillary light reflex; A demonstration by means of pupillometer. Acta. Soc. Ophthalmol. Jpn. 93:955-941.

ENVIRONMENTAL TOXICOLOGY AND THE AGING VISUAL SYSTEM

Thomas J. Harbin

Department of Psychology
University of North Carolina at Greensboro
Greensboro, NC 27412

Research on the environmental toxicology of the visual system has virtually ignored the special concerns of the elderly. In this chapter, the elderly will be portrayed as an at-risk group for the effects of environmental toxicants upon changes in the eye and possible interactions with toxic substances present in the environment.

The contamination of the environment by toxic substances is a serious problem and is increasing at an alarming rate. The development of new chemicals for industrial, agricultural, and other uses is accelerating. In addition, the volume of toxic substances released into the air, earth, and water is increasing as the world population grows and as industrial processes and chemical-dependent agricultural practices are introduced into new geographical areas. The development of institutions, legislation, and procedures for the protection of the environment is lagging well behind the pace of contamination. Coordinated procedures for the registration, testing distribution, and utilization of new substances have yet to be formalized in most parts of the world. Similarly, procedures for treatment of hazardous substances, cleanup of existing areas of contamination, and identification of the sources of contamination are inconsistent and infrequently applied.

Although exposure to environmental toxicants is undesirable for anyone, there are groups for whom certain substances pose a particular hazard. For example, carbon monoxide exposure is especially hazardous to those with coronary artery disease, pulmonary disease, and others sensitive to disruption of oxygen intake and metabolism. Developing fetuses and small children are susceptible to a wide range of toxicants, due to the rapid anatomical and physiological changes occurring at the age.

FORMAL AND INFORMAL TESTING PROCEDURES NOW PLACE SPECIAL EMPHASIS UPON THESE GROUPS AND OTHERS

The elderly represent a large and identifiable group of persons who may be especially vulnerable to environmental pollutants. Since the beginning of the 20th century, the number of individuals age 65 years or older has increased rapidly: from 3 million persons comprising 3% of the U.S. population, in 1980. This "graying of America", due to increased fertility rates, increased infant survival rates, and significant improvements in nutrition and health care (e.g., the development of vaccines and antibiotics), is expected to continue for the next several decades. By the year 2025, 40 million persons, comprising over 20% of the U.S. population, will be 65 years of age or older.

In the aged, there is an increasing vulnerability to all stressors, including those imposed by the environment. Age-associated changes in anatomy and physiological function place the elderly individual at risk for a variety of medical problems, exemplified by the disproportionately large percentage of individuals with vision problems. The same changes possibly render the elderly more susceptible to toxic effects upon vision following either acute or chronic exposure to potentially hazardous substances in the environment. Compounding this problem are those substances which pose no greater hazard for healthy older people than for younger persons but which may interact with age-related diseases. The dangers of carbon monoxide exposure for coronary heart disease patients illustrates this concern.

In addition to physiological changes, there are many social factors which increase the chances of toxicological insult in the elderly. The elderly tend to be located in areas which receive relatively high concentrations of pollutants. They are more likely to live in older neighborhoods which are closer to industrial centers. In recent years, the average age of farmers has been increasing and agricultural chemicals are a significant source of environmental pollution. In addition, the elderly are less educated than younger cohorts, a fact which has meant that they are more likely to have worked in hazardous labor situations and to have spent more time in such occupations. These factors render the elderly more susceptible to toxic substances which have delayed effects (e.g., asbestos) or which require an extended history of exposure to produce their effects (e.g., cotton dust), even though their effects may not be related to age per se.

Despite the overwhelming need for information regarding the interaction between age and the effects of environmental substances which are toxic to the visual system, virtually no research has been conducted on this topic. Though the importance of age has long been acknowledged for young organisms, it has yet to be realized that the rapid physiological changes occurring in the elderly may also be crucial to consider form a toxicological perspective. Fortunately, this fact has been recognized in the area of pharmaceutical drug development and testing. An important body of knowledge that has emerged from the action of drugs, quite apart from changes in the target organs or tissues. For example, while there are many alterations of visual system anatomy and physiology which render the older eye vulnerable to the toxic effects of therapeutic drugs, changes in drug uptake, distribution, and clearance often exacerbate these effects. Alterations in the vascular system, metabolism and renal and hepatic function have proven crucial to the understanding of the actions of pharmaceutical drugs in the elderly. The same will undoubtedly prove to be the case with environmental toxicants.

As result of the paucity of information available regarding possible interactions between adult age and visual system toxicants, this chapter will discuss changes in the anatomy, physiology , and function of the visual system with age and propose likely interactions with visual systems toxicants. Potential interactions will thus be indicated for future research. The identification of potentially hazardous substances for the elderly visual system will rely heavily upon Grant (1974) and the Merck index (Windholz, 1983).

The elderly eye often undergoes a discoloration of the lens. This is primarily characterized as a yellowing and opacification of the tissue (Weale, 1963) which reduces the amount of light reaching the receptors. In addition to increasing intensity thresholds, this discoloration will also alter the spectral characteristics of light reaching the retina. The primary agents causing lens discolorations are the metals copper, mercury, and silver. Exposure to copper compounds occurs primarily in smelting operations and in electroplating. The toxic effects of organic mercury compounds has been heavily investigated, primarily due to the catastrophic exposure suffered by the inhabitants of the Minamata area of Japan (Takeuchi, 1968). In addition to other serious visual system effects, organic mercurials produce lens discoloration. Systemic silver exposure leads to a characteristic alteration of the lens referred to as argyrosis, and often occurs upon exposure to photographic materials.

In addition to the lens, discoloration of the cornea seems to be a concomitant of advancing age. Toxicant-induced corneal discolorations has been reported from exposure to hydroquinone, (a photographic developer). Exposure typically occurs during the manufacturing process or in photographic laboratories. Exposed workers develop a brownish tinge overlaying both the cornea and the sclera (Anderson, 1947).

One of the most notable age changes in the light-conducting properties of the eye is cataract formation. These opacities are usually progressive and often result in the need for ocular surgery. A wide range of chemicals have been found to produce cataracts in humans, including deferoxamine (an iron chelating agent), dinitro-o-cresol (used as a herbicide and insecticide), dinitrophenol (used in the manufacture of many dyes), naphthols (also used in manufacturing dyes as well as synthetic perfumes), nitrocellulose paints and dyes, and trinitrotoluene (TNT, an explosive).

Animal research has identified many more cataract-inducing substances, including cobaltous chloride (used in meterological gauges, electroplating, paint manufacture, and in agricultural feed and fertilizer), decahydronaphthalene (a solvent), 2,6-dichloro-4-nitroaniline (applied to fruits and vegetables to inhibit mold), Diquat (an herbicide similar to Paraquat), naphthalene (used as a moth repellant and insecticide), tetrahydronaphthalene (a degreasing agent and solvent), thallium (an insecticide and rodenticide), and tretamine (used in the manufacture of resinous materials and as a finishing agent in textile production).

The elderly also experience a thickening of the lens and a loss of elasticity, reducing the ability of the lens to accommodate. There are myriad environmental toxicants which interfere with accommodation. Their primary mode of operation is to interfere with the action of the peripheral cholinergic nervous system. The most widespread example of the class of pollutant is the cholinesterase-inhibiting pesticides, such as Malathion, Parathion, Sevin, Niphos, and Dipterex. Exposure to the ciliary muscle with resulting loss of accommodative flexibility and spasm for near accommodation.

Age-related changes in vascular status, such as general arteriosclerosis and diabetic vascular disease can have severely detrimental repercussions for vision. The resultant degeneration of the retinal vasculature can seriously impair or eliminate vision. This pathology could potentially interact with substances which have been found to produce retinal hemorrhages. Examples of these include inorganic arsenic compounds (used in the smelting of ores, as insecticides, and as wood preservatives), benzene (formerly used widely as a solvent), lead, methyl bromide (a fire extinguisher and fumigant), trichloroethylene (used as a solvent and in dry cleaning), and Warfarin (a rodenticide).

Both Ordy and Brizzee (1979) and Kline and Schieber (1985) have noted that there are alterations in the retina with age. One such alteration is the atrophy of retinal ganglion cells, often accompanied by decreases in dendritic arborization. Several substances have been shown to damage ganglion cells, but as Grant (1974) indicated, research has failed to determine whether this damage is a primary effect of the toxicants or secondary to other changes such as elevated intracranial and intraocular pressure, or optic neuritis. Substances implicated include thallium, methanol (very widely used as a solvent and in the manufacture of many products such as animal and vegetable oils, plastics, and pharmaceutical preparations), and carbon disulphide (used in the manufacture or rayon, carbon tetrachloride, soil disinfectants, resins, and rubbers).

The posterior chamber of the eye is also subject to changes with age. There is a shrinkage of the vitreous in many older individuals producing, among many others changes, retinal individuals producing, among many other changes, retinal detachment. There are several toxicants with potential for interaction with this age change, thus exacerbating the tendency toward retinal detachment (often accompanied by choroidal edema and degeneration). The most commonly encountered are deferoxamine and naphthalene.

The elderly often undergo a degeneration of photo receptors, especially in the macula. Receptor damage, often in the form of interference with pigmentation (and sometimes producing a central scotoma), can be exacerbated by systemic exposure to copper and mercuric chloride (used in the manufacture of papers and paints, and as a fungicide and pesticide).

The incidence of glaucoma, with its associated elevation of intraocular pressure, increases with advancing age and is responsible for severe disturbances of vision in many older individuals. The disease is especially insidious because the patient experiences few symptoms until considerable ocular damage has occurred. The primary environmental toxicant resulting in elevated intraocular pressure is tetraethyl lead, an anti-knock compound added to gasoline.

In addition to demonstrable physiological and anatomical changes, the functional integrity of the visual system declines with age (Kline and Schieber, 1985). One such change involves a restriction in the functional sized of the visual field. Many substances have been shown to cause restriction of the visual fields. These include carbon monoxide, carbon tetrachloride, chlorodinitrobenzene (used in the manufacture of picric acid and certain explosives), ethylmercuritoluenesulfonanilide (a fungicide), methanol, methyl bromide, naphthalene, and trichloroethylene.

One area of environmental toxicology which has devoted some research to the issue of age differences in susceptibility is the investigation of potential neurotoxicological effects of carbon monoxide (CO). CO is a by-product of incomplete combustion and is a prominent constituent of the atmospheres of industrialized societies. The major sources of exposure to CO are the burning of fossil fuels in automobiles, industry, and home heating systems, and cigarette smoking.

The primary toxic effect of CO is due to its affinity for hemoglobin, which, in humans, is approximately 200 times greater than the affinity of oxygen for hemoglobin. Therefore, many of the physiological effects of CO exposure are similar to those of the comparative effects of CO-induced hypoxia (Goldsmith & Landaw, 1968). There are discrepancies in the comparative effects of CO-induced hypoxia and hypoxic hypoxia, however, and secondary toxic effects are possible (Benignus et al., 1983).

The elderly may be at particular risk for the effects of CO upon neurobehavioral function for several reasons. First, perceptual-motor performance is generally deficient in the elderly, relative to younger people (Salthous, 1985). Therefore, further degradation due to CO could be potentially more functionally significant for the elderly. Second, the perceptual-motor processes of the elderly are more susceptible to disruption from a number of sources (e.g., from increases in task complexity; Cerella et al., 1980). Finally, the elderly may be at greater risk for CO-induced hypoxia. The elderly have lower rates of brain oxygen consumption (Dastur et al., 1963) and they may also be marginally hypoxic due to arterial degeneration and other circulatory changes. In addition, the elderly lung is deficient in its ability to diffuse gases (Bates et al., 1971) and this may serve to exacerbate the toxic effects of CO.

Groll-Knapp et al. (1982) measured click evoked potentials during sleep in young and elderly subjects at a COHb level of approximately 8%. They reported generally increased amplitudes after CO exposure, especially for earlier components. There were no pronounced age differences in the CO effects. Hosko (1970) reported no changes in the flash evoked potential for COHb concentrations up to 14%. Above 20%, early components of the evoked potential increased in amplitude. A study by Harbin Benignus, Muller, and Barton (1988) investigated the effects of low-level CO exposure (5% COHb) upon the late portion of the visual evoked potential in young and elderly men. This component of the evoked potential, known as P300, has been linked to various aspects of cognitive behavior (Hillyard and Kutas, 1983). Results indicated no effects of CO upon either the amplitude or latency of the P3000 in either group. Young subjects in this study absorbed more CO than the elderly. This may be the result of a paradoxically advantageous decline in the respiratory function with age. Cohen (1964) reported that the elderly have decreased pulmonary diffusing capacity, a finding replicated by Georges et al. (1978). Changes in membrane permeability with age have also been reported (Bates et al., 1971). Stupfel et al. (1975) found that senescent rats (590 to 700 days) were less subject to lethal hypoxia than younger rats. These findings indicate that, in addition to absorbing CO less readily than younger subjects, the elderly may also be less susceptible to its hypoxic effects.

CONCLUDING REMARKS

In the chapter, no attempt has been made to list all of the possible substances which could potentially interact with changes in the visual system due to age or age-related disease. Only those substances which act systemically upon the visual system have been reviewed, and of those, consideration has been limited to toxicants which are either representative of classes of toxicants or are common constituents of the working or living environments of the United States. There are literally thousands of chemicals that are toxic when applied directly to the eyes. These are encountered in the form of dusts or from accidental splashes.

The actions of most environmental toxicants are difficult to elucidate due to the fact that industrial agricultural chemicals are usually combinations of many potentially hazardous substances, formulas are protected, and quality control and documentation are not as rigorous as with medicinal compounds. This results in incomplete knowledge regarding which constituents of a commercial product produce the observed injury. However, detailed discussion of some specific visual system toxicants can be found in Merigan and Weis (1980). Pharmaceutical substances have also not been reviewed in this chapter. Information regarding possible toxic effects of these drugs can be obtained in Grant (1974 and Merck index (Windholz, 1983).

With the exception of the developing fetus and young children, there is probably no larger group in the United States that can be considered more susceptible to the effects of environmental toxicants upon the visual system than the elderly. The virtual absence of research in this area is disturbing. In view of the pervasiveness of visual deficits in the elderly, and the enormous cost both in monetary terms and in terms of human suffering, it is to be hoped that this state of affairs will be rectified in the future.

REFERENCES

Anderson, B. (19747). Corneal and conjenctival pigmentation among workers engaged in manufacture of hydroquinone. Archives of Opthalmology,38, 812-826.

Bates, D.V., Macklen, P.T., & Christie, R.V. (1971). Respiratory Function in Diesease: 2nd edition. Philadelphia: W.B. Saunders, pp. 75-92.

Benignus, V.A., Muller, K.E., Barton, C.N., Prah, J.D., & Benignus, L.L. (1983). Neurobehavioral effects of carbon monoxide (CO) exposure in humans. U.S. Army Medical Research and Development Command,Fort Detrick, Frederick, MD, 21701.

Cerella, J., Poon, L.W., & Williams, D.M. (1980). Age and the complexity hypothesis. In: Aging in the 1980's: Psychological Issues, edited by L.W. Poon. Washington, DC: American Psychological Association, pp. 125-134.

Cohen, J.E. (1964). Age and the pulmonary diffusing capacity. In L. Cander & J. Moyer (Eds.), Aging of the lung. New York: Grune & Stratton, pp. 163-172.

Dastur, D.K., Lane, M.H., Hansen, D.B., Kety, S.S., Butler, R.N., Perlin, S., & Sokoloff, L. (1963). Effects of aging on cerebral circulation and metabolism in man. In: Human Aging: A Biological and Behavioral Study. Washington, DC: U.S. Government Printing Office (USPHS Publication 986), pp. 57-76.

Georges, R., Saumon, G., & Loiseau, A. (1978). The relationship of age to pulmonary membrane conductance and capillary blood volume. American Review of Respiratory Disease, 117, 1069-1078.

Goldsmith, J.R. & Landaw, S.A. (1968). Carbon monoxide and human health. Science 162: 1352-1359.

Grant, .M. (1974). Toxicology of the eye. Springfield, IL: Charles C. Thomas.

Groll-Knapp, E., Haider, M., Jenkner, H., Liebich, H., Neuberger, M., & Trimmel, M. (1982). Moderate carbon monoxide exposure during sleep: Neuro - and psychophysiological effects in young and elderly people. Neurobehavioral Toxicology and Teratology, 4, 709-716.

Harbin, T.J., Benignus, V.A., Muller, K.A., & Barton, C.N. (1988). The effects of low-level carbon monoxide exposure upon evoked cortical potentials in young and elderly men. Neurotoxicology and Teratology, 10, 93-100

I thank Lynn Newlin-Clapp for technical assistance and Marian E. Saffer for thoughtful comments on this manuscript.

UPDATE 1990

It would be gratifying to report that the field of environmental toxicology has evidenced an increased emphasis upon the elderly since the symposium reported in this volume was held. Unfortunately, this is not the case, particularly as concerns the effects of environmental toxicants upon he aging visual system. A review of the toxicological literature for the past five years developmental perspective. These investigations usually fall into one of three categories; 1) case reports of accidentalover-exposure of an elderly person to a toxicant or therapeutic drug (e.g., Fladelius et al., 1987), 2) investigations of samples of subjects with a large age range, but with little or no analysis of the age variable (e.g., Ludolph et al., 1986), or 3) investigations of interactions among age, damenting illness, and alcohol abuse upon visual system integrity (e.g., St. Clair at al., 1985). A more systematic program of toxicological research from an adult-developmental perspective is warranted if the special vulnerabilities of the elderly to visual system toxicants is to be understood. Such understanding will be crucial to the responsible use and regulation of visual system toxicants and the maintenance of adequate vision in old age.

References

Fladelius, H.C., Petrera, J.E., Skjodt, K., & Trajaborg, W., 1987, Ocular ethambutol toxicity. A case report with electrophysiological considerations and a review of Danish cases. Acta Opthalmologica. 65, 251-255.

Ludolph, A., Elger, C.E. Sennhenn, R., & Bertram, H.P., 1986, Chronic thallium exposure in cement plant workers: Clinical and electrophysiological data. Trace Elements in Medicine, 3, 121-125.

St. Clair, D.M., Blackwood, D.H., & Christie, J.E., 1985, P3 and other long latency auditory evoked potentials in presenile damentia Alzheimer type and alcoholic Korsakoff syndrome. Br. J. Psychiatry, 147, 702-706.

CONTRIBUTORS

Donald Armstrong, Ph.D., Dr.Sc.
Professor and Chairman
Department of Medical Technology and
Program in Nuclear Medicine Technology
School of Health Related Professions
Faculty of Health Sciences
UB Clinical Center
State University of New York
Buffalo, New York

Marcus D. Benedetto, Ph.D.
Director, Ophthalmic Research Center
Orlando, Florida

Amos Ben-Zvi, M.D.
Chief, Dept. of Pediatric Immunology
Hadassah University
School of Medicine
Jerusalem, Israel

M.L. Chandler, Ph.D.
Associate Technical Director
Alcon Laboratories, Inc.
Fort Worth, Texas

Louis DeSantis, Ph.D.
Sr. Director Alcon Laboratories, Inc.
Fort Worth, Texas

William P. Dunlap, Ph.D.
Professor, Departments of
Psychology and Biostatistics
Tulane University
New Orleans, Louisiana

Robert E. Dustman, Ph.D.
Professor and Director
Research Career Scientist
Neuropsychology Laboratory
Veterans Administration Medical Ctr.
Salt Lake City, Utah

Craig Eldred, Ph.D.
Research Associate Professor
Mason Institute of Ophthalmology
Dept. of Opthalmology School of Medicine
University of Missouri - Columbia
Columbia, Missouri

Rita Y. Emmerson, Ph.D.
Assistant Research Professor
Neuropsychology Laboratory
Veterans Administration Medical Ctr.
Salt Lake City, Utah

Leslie S. Fujikawa, M.D.
Assistant Professor
Department of Ophthalmology
Pacific Medical Center
2340 Clay Street
San Francisco, California

Brenda W. Griffin, Ph.D.
Assistant Technical Director
Alcon Laboratories, Inc.
Fort Worth, Texas

Thomas J. Harbin, Ph.D.
Licensed Practicing Psychologist
Gullick and Associates
Fayetteville, North Carolina

Jessica Jahngen-Hodge, B.A
Research Associate
Laboratory for Nutrition and
Vision Research
United States Dept. of Agriculture
Human Nutrition Research Center
on Aging
Tufts University
Boston, Massachusetts

Betty M. Johnson, M.S.
Research Assistant
Department of Ophthalmology and
Neurosurgery
Estelle Doheny Eye Institute
University of Southern California
Los Angeles, California

Satoshi Ishikawa, M.D.
Professor and Chairman
Department of Opthalmology
School of Medicine
Kitasato University
Sagamihara, Kanagawa, Japan

Martin L. Katz, Ph.D.
Research Associate Professor
Mason Institute of Ophthalmology
Department of Ophthalmology
School of Medicine
University of Missouri - Columbia
Columbia, Missouri

Robert Kennedy, Ph.D.
Vice Pres. for Research & Development
Essex Corporation
Orlando, Florida

Michael F. Marmor, M.D.
Professor and Chairman
Department of Ophthalmology
Stanford University Medical Center
Stanford, California

Larry G. Meyer, M.S.
Research Physiologist
Environmental Physiology Division
Naval Aerospace Medical Research Lab.
Naval Air Station
Pensacola, Florida

Michael Miao, B.S.
Laboratory Assistant, Departments of
Opthalmology and Neurosurgery
Estelle Doheny Eye Institute
University of Southern California
Los Angeles, California

J. Mark Ordy, Ph.D.
Senior Principal Investigator
Department of Biology
Fisons Pharmaceutical Corporation
Divisional Research and Development
Rochester, New York

Albert M. Prestrude, Ph.D.
Associate Professor and Director
Applied Experimental Psychology
Program
Department of Psychology
Virginia Polytechnic Institute
Blacksburg, Virginia

Sankaran Rajagopalan, M.D.
Research Associate
Lab. of Eye Pathology and Cell Biology
Department of Ophthalmology
School of Medicine
University of Maryland at Baltimore
Baltimore, Maryland

W. Gerald Robison, Jr., Ph.D.
Head, Section of Pathophysiology
Lab. of Mechanism of Ocular Diseases
National Eye Institute
National Institutes of Health
Bethesda, Maryland

Merlyn Rodrigues, M.D., Ph.D.
Professor and Director
Lab. of Eye Pathology and Cell Biology
Department of Ophthalmology
School of Medicine
University of Maryland at Baltimore
Baltimore, Maryland

Alfred A. Sadun, M.D., Ph.D.
Prof. Departments of
Ophthalmology and Neurosurgery
Estelle Doheny Eye Institute
University of Southern California
Los Angeles, California

Donald E. Shearer
Biological Technician
Neuropsychology Laboratory
Veterans Administration Med. Center
Salt Lake City, Utah

William Sleight, O.D.
Director
Center for Aging and Vision
New England College of Optometry
Boston, Massachusetts

Allen Taylor, Ph.D.
Associate Professor and Director
Lab. for Nutrition and Vision Research
United States Dept. of Agriculture
Human Nutriton Research Ctr. on Aging
Tufts University
Boston, Massachusetts

David Walsh, Ph.D.
Associate Prof., Dept. of Psychology
and Senior Research Associate
Andrus Gerontology Center
University of Southern California
University Park Campus
Los Angeles, California

Thomas M. Wengenack, Ph.D.
Post-Doctoral Fellow
Dept. of Neurobiology and Anatomy
University of Rochester
Rochester, New York

B.M. York, Ph.D.
Vice President of Research
Alcon Laboratories, Inc.
Fort Worth, Texas

INDEX